Australian High Country Owls

Dedication

For Deana Withey, Tressa Bloss and Carol Person, all heroes.
And for Deb who prevailed.

Australian High Country Owls

Jerry Olsen

CSIRO PUBLISHING

National Library of Australia Cataloguing-in-Publication entry

Olsen, Jerry.

Australian high country owls/by Jerry Olsen.

9780643097056 (pbk.)
9780643104105 (epdf)
9780643104112 (epub)

Includes bibliographical references and index.

Owls – Australia.
Owls – Australia – Identification.
Mountain animals – Australia.

598.970994

Published by

CSIRO PUBLISHING
36 Gardiner Road, Clayton VIC 3168
Private Bag 10, Clayton South VIC 3169
Australia

Telephone: [+613] 9545 8555
Local call: 1300 788 000 (Australia only)
Fax: +61 3 9662 7555
Email: csiropublishing@csiro.au
Web site: www.publishing.csiro.au

Front cover: Powerful Owl (courtesy of David Stowe Photography)

Maps, photos and line drawings edited by Alan Nicol.
Line drawings by Nic Day and Robert Bartos.
Photos by Jerry Olsen unless otherwise noted.

Set in Adobe Minion Pro 11/13.5 and Adobe Helvetica Neue LT Std
Edited by Anne Findlay, Editing Works Pty Ltd
Cover design by Andrew Weatherill
Text design by James Kelly
Typeset by Desktop Concepts Pty Ltd, Melbourne
Index by Bruce Gillespie
Printed by Ingram Lightning Source

Jan26_RP_ILS

Contents

Preface and acknowledgements

Australia is the biogeographic centre for two major owl groups – the *Tyto* and the *Ninox*. Biologists outside of this region know little about Australian owls, and these owls differ in certain respects from their counterparts in North America and Eurasia. Australian biologists draw heavily upon overseas owl research when designing survey protocols and interpreting data, but how similar are the habitat types and calling behaviour of Australian and Northern Hemisphere owls? Is it harder to survey and study owls in Australia than in North America or Eurasia? Do Australian owls fill as many ecological niches with their two genera, *Ninox* and *Tyto*, as European or Canadian owls fill with many more genera – *Bubo*, *Tyto*, *Otus*, *Strix*, *Surnia*, *Glaucidium* and more? Does Australia have an ecological equivalent of the ubiquitous Great Horned Owl in North and South America, or of the Tawny Owl in Eurasia? Does Australia have small insectivorous owls such as the tiny Flammulated Owl or the sparrow-sized Elf Owl in North America? Do owls in Europe, Africa, South and North America roost on prey during the day as some Australian owls do?

The purpose of this book is to investigate Australian owls, in particular Australian *Ninox*, and the similarities and differences between these Australian owls and their counterparts elsewhere. The book discusses some past and recent research on Australian *Ninox* and *Tyto*, and contextualises these owl species with some well-known owls such as the Great Horned, Great Grey, Snowy, Little, Eagle, Tawny, Spotted, Northern Hawk, and Long-eared Owl. The book also discusses conservation techniques, and the conservation status of each Australian owl species and, in Appendix A, updates information on their biology. Appendix B gives guidelines for rehabilitating injured and orphaned owls.

I would like to thank Susan Trost, Stephen Debus, Rod Kavanagh, Ian Fraser, Jim Duncan, Tracy Fleming and Jim Hone for reviewing parts of earlier drafts. Thanks to numerous people for field assistance and advice including Tony Tucker, Nancy Fitzsimmons, Tom and Gillian Bellas, Jeanine Randall, Susan Meyers, Greg Hayes, Dalice Trost, David Mallinson, Mark Osgood, Colin McClaren, and Frank and Tanya Barnes. ACT Parks and Conservation gave access to ACT Nature Parks and Namadgi National Park, and I am especially grateful to Michael Maconachie, Odile Arman, Steve Taylor, Christie Gould, Murray Evans, Brett McNamara, Paul Higginbotham, Kate Boyd, Meg Doepel, Simon Tozer and Darren Rosso who assisted in the field. David Drynan and Lisa Hardy of the Australian Bird and Bat Banding Scheme supplied ABBBS bands and Vic Hurley generously supplied colour-bands for some of the Boobook studies. Thanks to David Judge for statistical advice and to Greg Flowers, Joe McAuliffe and Robyn Lawrence for help with the ACT Powerful Owl study. Rod Kavanagh, Ed McNabb, Raylene Cooke, Fiona Hogan, Todd Soderquist, Chris Pavey, Paul Peake, Tony Rose, Bill Mannan, Arthur Georges, Jim Hone, Charles Krebs, Stephen Sarre, Terry Dennis, Jeff Jolly, Nick Mooney, David Bird, Marcus Moleón, Erkki Korpimäki, Jesus Lemus, Joan Real, Luis Palma, Michael Wink, Iñigo Zuberogoitia, Lander Astorkia, Inaki Castillo, Claus and Ingrid König, Dick Schodde, Ian Mason, Robert Palmer, Joni Downs, Denver Holt, Erik Powell, Kate Davis, David Anderson, Bud Anderson, Jonny Groves, Lloyd Kiff, Ray Cromie, Allan Fish, Jeff Marks, Geoff Holroyd, Gordon Court, Will Osborne, David Williams, Walter Boles, Sue Briggs, Richard Norris, Karen Mow, Soliana Tedla, Jo Shevchenko, Kerrie Aust and Les Boyd provided logistic help and helpful discussion. Members of the Canberra Ornithologists group, including Geoffrey Dabb, Anthony Overs, Michael Lenz, Susan Robertson, Chris Davey, Barbara Allan, Richard Allan and Mark Clayton, provided invaluable help with Canberra Boobooks and Powerful Owls. Thanks also to CSIRO Publishing especially John Manger, Deepa Travers, Anne Findlay and Tracey Millen for their professional advice and assistance throughout the process of publishing this book.

Particular thanks to Alan Nicol for skilled editing of photos and drawings used throughout the book, and for maps used in Appendix A. Thanks also to Robert Bartos for his detailed drawings used in Appendix B, and to Nic Day for his expert drawings used in the main text.

1

What is an owl?

The Hobby was bent on killing or driving away some enemy. She arrowed down through the canopy of a stringybark with her wings in a tight arc, then looped up, rolled over, and stooped through the branches again, calling as she attacked. A Southern Boobook and its fluffy young were day-roosting in Canberra woodland. The young had become restless and started to stretch and move in a shady tree, too close to the Hobby's nest. Even though Boobooks are hardly bigger than Hobbies, something in the owl's form and behaviour singled it out as a natural enemy (see Figures 1.1, 1.2 and 1.3). Many birds react this way to owls, and researchers around the world tether owls to perches in front of nets to capture falcons and other birds of prey that attack them and try to drive them away.

What makes an owl so recognisable to birds such as Hobbies, and recognisable to us? Compared to most other birds, owls are relatively large-headed with relatively large brains and very large forward-looking eyes, so large that their eyes are virtually immobile in their sockets. Most other birds and animals have smaller sideways-looking eyes that can look out for predators attacking from the side or behind. The large size of an owl's eyes gives them excellent vision at night, similar to a domestic cat's vision, and like a cat they can also see well during daytime. This large head with forward-looking eyes gives them an almost human look and makes them appealing to humans in the same way that cats or monkeys are appealing, especially their young. Owls have a rather short, decurved, powerful beak that is partially hidden in rictal bristles, almost like a bristly moustache. Some, like the Boobook, have white 'eyebrows' and all owls have closable upper eyelids that add to this 'human' look. They sleep head-erect with eyes closed by twin eyelids. Probably

Figure 1.1 Adult Southern Boobook feeding its fledged young. The adult closes its eyes to avoid damage from the juvenile's sharp beak and flapping wings.

Figure 1.2 Owls have a characteristic head and body shape that makes them recognisable to other birds.

Figure 1.3 Birds immediately attack owls as enemies. This Eagle Owl is tethered in front of a mist net by researchers Iñigo Zuberogoitia and Lander Astorkia to capture Peregrine Falcons in Spain.

their heads are too big to tuck under their wings, as falcons and other birds do. Owls have a well-developed nictitating membrane, a transparent fold of skin that can be drawn across as a third eyelid. They activate this eyelid when struggling prey threatens to injure the owl's eyes, or when the owl is feeding prey to their own young, apparently because they fear getting poked by an outstretched flapping wing or a sharp beak.

This characteristic head shape with large eyes leads to behaviours typical of the group. As with humans, owls' eyes give stereoscopic perception, probably more so than in any other bird. Unlike humans, their eyes have almost no movement in their sockets due to the fusing of the sclerotic ring with the skull. Owls' eyes are simply too big to move in their sockets and their flattened faces allow their eyes to be spaced far apart (see Figure 1.4). This helps them gauge enemies and prey three-dimensionally. By circling and bobbing its head, an owl sights an object from a variety of minutely different angles and receives a better picture of it; because of this, they can make better decisions about hunting in thick cover, or catching prey in mid-air.

This young owl, sitting upright on a jagged limb and calling steadily, began to 'head-wind', a behaviour peculiar to owls (shown in Figure 1.5). When owls look

Figure 1.4 The eyes of Southern Boobooks and other owls are too large to allow much movement in their sockets.

Figure 1.5 Young owls, such as Boobooks, head-wind so they can better see an object. Adults tend not to do this.

hard at an object, in this case looking hard at us humans, they sometimes wind their flattened faces in a plane, like a dinner plate moving in circles, first clockwise, then back. Young owls head-wind so they can better see an object and bring it into focus at their points of retinal reception (the fovea). They wiggle their whole bodies back and forth, wagging their tails when they head-wind.

Because of the fusing of the sclerotic ring with the skull, owls need to swivel their heads to look in another direction. Though owls look short-necked, they have 14 cervical vertebrae that allow them to easily turn and look over their shoulder, even swivel their heads 270°. Young owls like to turn their heads upside down to gain a different perspective. As with this young owl we were watching, the young Hobbies in the nest nearby have almost no movement in their eye sockets either, and they too can turn their heads upside down to watch butterflies passing during the day, and look backwards over their shoulders. But Hobbies don't head-wind as Boobooks do.

This adult and juvenile sitting in the shady tree were, like most owls, coloured in sombre tones, cryptically patterned for daytime camouflage, in much the same way as frogmouths and nightjars. In fact, the two owls we had been watching perched all day close to the Hobbies' nest and the Hobbies had not attacked them. Only when dusk approached and the owls became restless, began to move on their perches, stretch and look around, did they draw attention from other birds that attacked them as enemies.

Like most birds, owls have a lightweight skeleton for flight, and their wing-bones are relatively long, associated with a broad surface area, which gives them a low wing loading (shown in Figure 1.6). They can easily lift and carry heavy prey, even in still air, something that Hobbies or Peregrines with a higher wing loading and narrow wings can't easily do; falcons need a head wind to lift heavy prey.

Though the owl's skeleton is lightweight, the foot bones and tarsometatarsi of owls are relatively short and stout, with highly developed curved talons on all four toes for efficient killing and carrying of large prey. Some predators, such as the Tawny Frogmouth nesting nearby in these same woods, do not have these dangerous toes and talons (shown in Figure 1.7). Tawny Frogmouths kill with their beaks, not their feet. Researchers handling owls and other birds of prey, worry more about an owl's powerful feet than about their beaks, though both can cause injury.

Most owls differ from most diurnal birds of prey, like falcons, in that their legs are feathered down to the toes. Some researchers say this is for protection against victims that bite, such as rats, but small owls that catch insects also have feathered legs, and most diurnal raptors, like goshawks, sea-eagles or falcons, do not need feathered legs even though they take and subdue large, dangerous prey.

The biggest difference between the feet of owls and those of diurnal raptors (except Ospreys) is that owls have a 'reversible' outer toe. They can perch with three

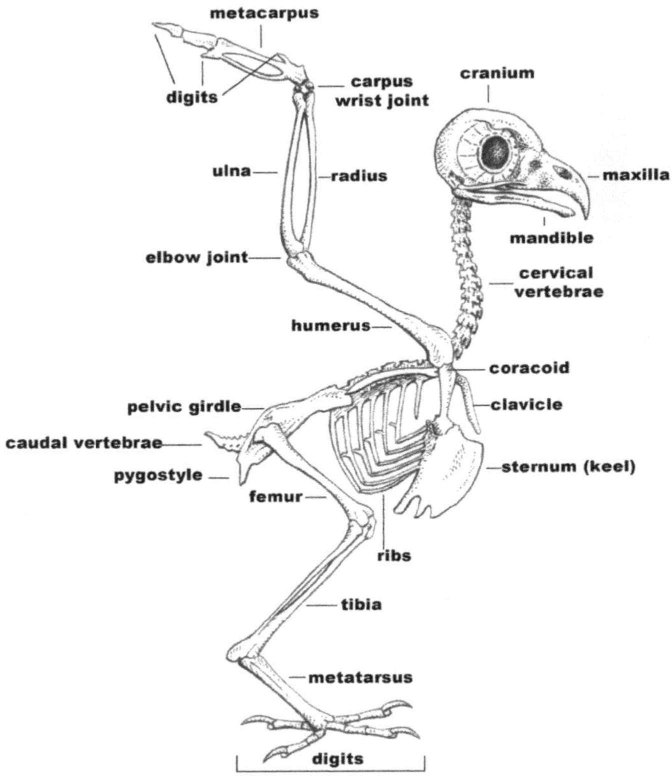

Figure 1.6 Typical owl skeleton.

Figure 1.7 Unlike Southern Boobooks, Tawny Frogmouths are not owls. They kill with their beaks, not their feet.

Figure 1.8 Owls can have three toes pointing forward, or swing the outer toe behind them and perch or catch prey with two toes pointing forward and two facing backwards, a 'zygodactyl' arrangement.

toes pointing forward and one pointing back, the usual 'anisodactyl' manner of most birds, including falcons and eagles. Or they can swing the outer toe behind them and perch or catch prey with two toes pointing forward and two facing backwards, a 'zygodactyl' arrangement (see Figures 1.8 and 1.9). This provides a wider, more balanced spread. Because this 2 × 2 arrangement is almost always used when an owl reaches for or carries prey (small prey is often carried in the bill), some have argued that owls have evolved this special advantage over other raptors to better clutch and carry heavy prey. It is equally likely, however, that owls are taxonomically close to birds like woodpeckers that already had two toes forward and two toes backward; the 2 × 2 toes arrangement probably evolved first and the hunting advantage for owls evolved second, if it truly is an advantage. Again, all diurnal raptors except Ospreys get along fine without it.

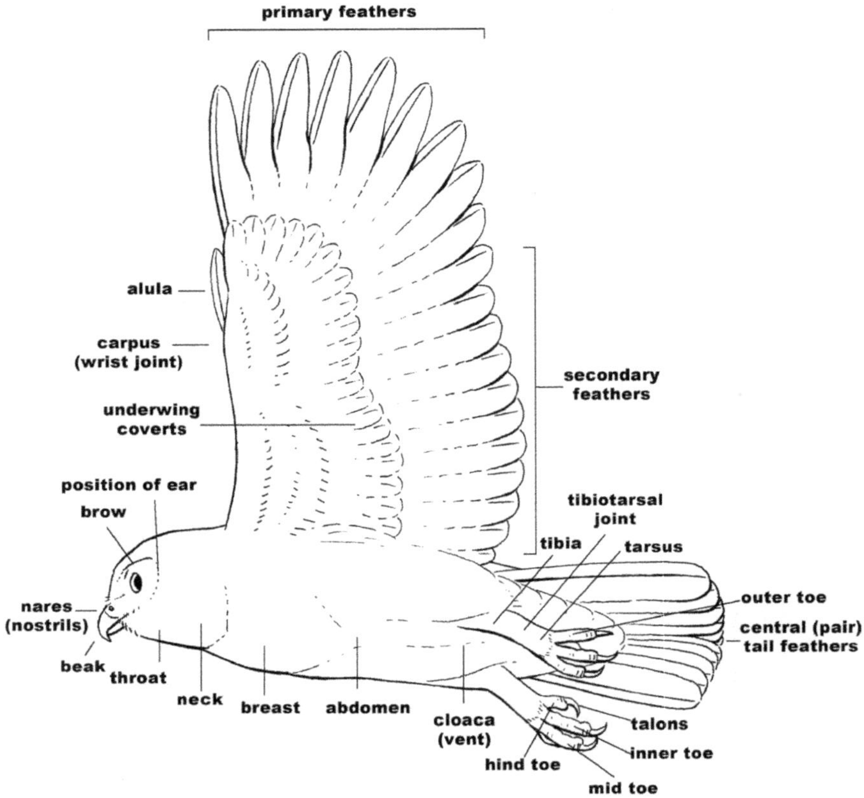

Figure 1.9 Body parts of a Powerful Owl.

The adult Boobook continued to forage in moonlight over the open field, catching Christmas Beetles and the occasional moth with its toes wide and legs stretched forward. It brought these back in its beak to the trilling juvenile. The juvenile ate them, begged for more, and followed the hunting adult across the clearing to the edge of another woodland.

2

What is a *Ninox*?

To understand Australian hawk-owls we need to understand something about owl taxonomy. The world's owls, the Strigiformes, are divided into two families: (1) Tytonidae – 'bay owls' or 'barn owls' and (2) Strigidae – 'typical owls'. The Tytonidae has some 27 species distributed around the world, Strigidae has some 219 species, depending on taxonomic treatment. Tytonidae, like the Barn Owl species found around the world, with heart-shaped facial discs and no ear tufts, are familiar to many people (see Colour Plate 1 and Figure 2.1). Most are in one genus, *Tyto*. *Tyto* species differ in colour, from mostly white to almost black, as the Barn, Grass, Sooty, and Masked Owls do in Australia. All species have a characteristic head-swaying threat display that no Strigidae have. Tytonidae have 15 secondaries, Strigidae have 11 to 18. Tytonidae owls do not like the cold, you do not find them in the Arctic. They tend to be most common in the tropics and subtropics and few species breed further than 40 degrees north or south of the equator.

Strigidae owls breed from the Arctic to the sub-Antarctic, and they vary much more in size, colour and shape than do Tytonidae owls. The familiar cat-like 'hoot owl' in North America, the Great Horned Owl, as well as the Tawny Owl in England, Spotted Owl in western North America, Snowy Owl in the Arctic, long-legged Burrowing Owl in North and South America, tiny ear-tufted Screech and Scops Owls and the non-ear-tufted Pygmy Owls across the world (see Colour Plate 2), and the Boobook, Barking, Powerful and Rufous Owls in Australia are all Strigidae (as in Figure 2.2). Unlike Tytonidae, that tend to have a variety of

Figure 2.1 Owls in the Family Tytonidae, such as this Barn Owl from Australia, have heart-shaped facial discs and no ear tufts.

Figure 2.2 Owls in the Family Strigidae vary much more in size, colour and shape than do Tytonidae owls. Some, like this Spotted Owl from the United States, or Southern Boobooks, have no ear tufts. Others, like the Eagle Owl, have ear tufts.

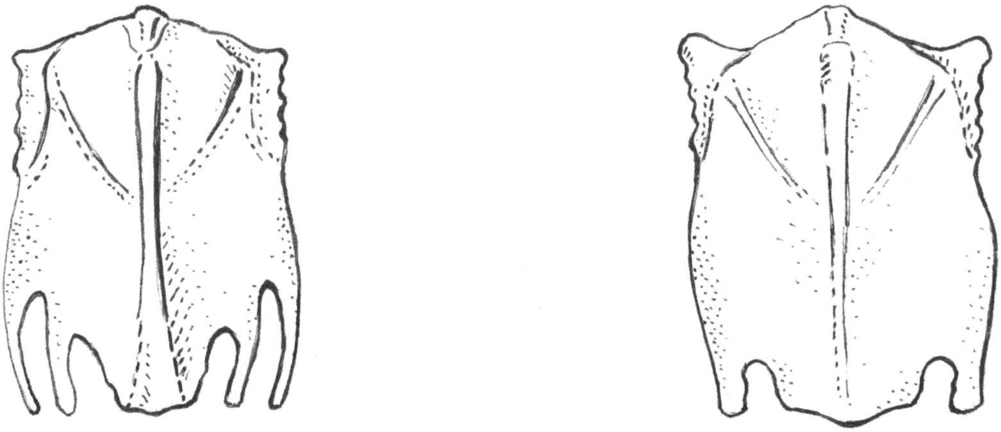

Figure 2.3 A typical Tytonidae (right) sternum has two shallow notes near the abdomen; a typical Strigidae sternum (left) has two deep emarginations on each side.

screeching calls, many Strigidae have a characteristic 'hooting' call, for example, the 'Boobook' call of Southern Boobooks.

Tytonidae owls have inner and central toes of about the same length, while Strigidae owls have an inner toe that is shorter than the middle toe. The claw of the central toe in Tytonidae is serrated on the underside, in Strigidae it is not. The skull of Tytonidae appears to be long and pinched, in Strigidae comparatively short. The shape of the sternum (breastbone) is different in the two families. In Tytonidae the sharp edge of the sternum, the carina, is broad and narrows towards the abdomen, whereas in Strigidae it is narrow at the upper part and broadens towards the abdomen. In Tytonidae the sternum has two shallow notches near the abdomen, in Strigidae there are two deep emarginations on each side (shown in Figure 2.3).

No Tytonidae have emarginated (notched) primaries, while many Strigidae, such as Boobooks, do. Tytonidae are rather long-legged owls with long wings and relatively short tails. The facial disc is more or less heart-shaped: the rim of feathers around the disc whose inner (central) edges run to the base of the bill form a narrow vertical ridge of two parallel lines (König and Weick 2008; Johnsgard 2002).

Tytonidae are said by some to be distinguished from Strigidae by a sharply defined facial disc, typically heart-shaped (Higgins 1999) but, to my eye, the differences are not so clear between some Tytonidae species like the two Sooty Owls in Australia and some Northern Hemisphere Strigidae owls such as the Boreal Owl. Even some individual Southern Boobooks have a fairly well defined facial disc. Many owls around the world, such as those in the genera *Bubo*, *Asio* and *Otus*, have ear tufts that make them look like cats, maybe because they evolved alongside cat species. No Australian owls have ear tufts.

Owl taxonomy

The genus is the first part of a taxonomic name. It denotes a category designed to include one or more species of plant or animal presumably of common phylogenetic origin. The genus is always italicised and capitalised in the first half of a binomial scientific name, for example, *Ninox rufa* for Rufous Owl. Taxonomy is often based on morphology, but the most recent analyses use DNA from blood, feathers or tissues. Wink *et al.* (2008, 2009) used the mitochondrial (mt) cytochrome *b* gene to study the finer details of speciation and phylogeny of the world's owls, and additionally sequenced the nuclear (nc) *RAG-1* gene for all groups of owls that were critical to a better understanding of relationships between owl genera and species. The family Tytonidae is said by some taxonomists to have only one genus, *Tyto*, but Wink *et al.* (2008, 2009) include *Phodilus*, the Bay Owls. In contrast, Strigidae is said to have some 25 genera, depending on taxonomic treatment. They list: *Otus*, *Psiloscops*, *Megascops*, *Pyrroglaux*, *Gymnoglaux*, *Ptilopsis*, *Mimizuku*, *Bubo*, *Pulsatrix*, *Strix*, *Jubula*, *Lophostrix*, *Surnia*, *Glaucidium*, *Taenioglaux*, *Xenoglaux*, *Athene*, *Micrathene?*, *Aegolius*, *Ninox*, *Uroglaux*, *Sceloglaux*, *Asio*, *Nesasio* and *Pseudoscops*.

Taxonomists also divide families into subfamilies, and often cluster genera that are similar into these subfamilies, and into 'tribes'. They identify these tribes by the suffix '-ini' (Table 2.1). DNA samples for all the world's owls are not complete, so Wink *et al.* (2008, 2009) made guesses about certain genera. *Pyrroglaux* and *Gymnoglaux* represent monotypic genera. *Pyrroglaux podarginus* has been described from the Palau Islands and *Gymnoglaux lawrencii* from Cuba. DNA analyses are required to see whether both taxa represent monotypic genera, and these are not included in Table 2.1. Two monotypic genera were included in the tribe Surniini, *Xenoglaux loweryi* from northern Peru and *Micrathene whitneyi* from south-western North America. Preliminary DNA sequence data only exist for *M. whitneyi*, which would place it outside the tribe Surniini (Wink *et al.* 2008) but close to the subfamily Surniinae. *Uroglaux dimorpha* (north-western New Guinea) and *Sceloglaux albifacies* (New Zealand) have been included in the tribe Ninoxini, which would make sense in view of distribution and general appearance.

Wink *et al.* did not fit the Australian *Ninox* into a tribe. They placed them in a subfamily Ninoxinae with two other genera because these Ninoxinae owls differed from owls in other tribes. Included were *Uroglaux*, with only one species, the Papuan Hawk-Owl, and *Sceloglaux*, the extinct New Zealand Laughing Owl. *Ninox* comprises some 25 owl species with Australasian distribution. According to their general appearance they could be related to the Surniinae and Athenini tribes that include the genera *Surnia*, the Northern Hawk-Owl; *Glaucidium*, the pygmy owls; and *Taenioglaux*, the small owlets once considered part of *Glaucidium*. In later

Table 2.1. Summary of subfamilies, tribes and genera of Strigidae owls based on monophyletic groups (after Wink *et al.* 2009)

Subfamily	Tribe	Genera
Striginae	Bubonini	*Bubo* (including the former *Nyctea, Ketupa* and *Scotopelia*)
	Strigini	*Strix* and *Jubula*
	Pulsatrigini	*Pulsatrix* and *Lophostrix*
	Megascopini	*Megascops* and *Psiloscops*
	Asionini	*Asio, Ptilopsis* (probably *Nesasio* and *Pseudoscops*)
	Otini	*Otus* and *Mimizuku*
Surniinae	Surniini	*Surnia, Glaucidium* and *Taenioglaux*
	Athenini	*Athene* (including former *Speotyto*) and *Micrathene* (?)
	Aegolini	*Aegolius*
Ninoxinae		*Ninox, Uroglaux* and *Sceloglaux*

chapters we consider some of these owl species, as well as species in *Bubo*, *Strix* and *Athene*, and compare these to *Ninox*.

The important point here is that Australia has only two owl genera: *Tyto* and *Ninox*. They fill niches often filled by a number of genera on islands and continents elsewhere in the world. Even small countries such as England have six breeding owl species across five genera, and islands such as Bali and Java have 11 breeding species across eight genera, depending on taxonomic treatment. The much larger Australian mainland has only two genera.

Tyto

The genus name *Tyto* is derived from the Greek for a type of owl, probably named for its cries. Australian *Tyto* can be divided into four groups: Sooty Owls, Masked Owls, Barn Owls and Grass Owls (Christidis and Boles 2008). The Barn and Grass Owls prefer open land; the Sooty and Masked Owls prefer forested or wooded land. All four groups are represented in Australia:

- Australian Barn Owl *Tyto delicatula*
- Australian Masked Owl *Tyto novaehollandiae*
- Tasmanian Masked Owl *Tyto castanops*
- Greater Sooty Owl *Tyto tenebricosa*
- Lesser Sooty Owl *Tyto multipunctata*
- Eastern Grass Owl *Tyto longimembris.*

There is still argument and confusion about how many *Tyto* species there are in Australia. According to the latest field guides for example, Graham Pizzey and Frank Knight (2007) *The Field Guide to the Birds of Australia* and the most recent official checklist of Australian birds, Christidis and Boles (2008) *Systematics and Taxonomy of Australian Birds,* the checklist that most Australian zoology journals require authors to use, Australia has four *Tyto* – the Barn Owl, Masked Owl, Sooty Owl and Grass Owl. But this is only an opinion. Dick Schodde and Ian Mason (1980) in *Nocturnal Birds of Australia* had separated the Sooty Owls into two species, the Sooty Owl and the Lesser Sooty Owl on the far north Queensland coast, and this was accepted for many years. Janette Norman and her colleagues (2002) argued that their DNA work showed the two Sooty Owls were only one species, as is currently accepted. David Hollands (2008) disagreed. In *Owls, Frogmouths and Nightjars of Australia* he said he had watched both Sooty Owls and Lesser Sooty Owls in the wild and they were surely two separate species. Michael Wink and his colleagues (2008) at the University of Heidelberg in Germany agreed. Their DNA analysis showed that Sooty Owls and Lesser Sooty Owls were separate species, as Schodde and Mason and Hollands had argued. Furthermore, Wink *et al.* argued that the Tasmanian Masked Owl and the mainland Masked Owl were also separate species. Their count then, for Australian *Tyto* species, was six, not four as is currently accepted in most Australian texts.

Ninox

Ninox comes from two words – Nisus and Noctua. In Greek mythology Nisus was the King of Megara. As he committed suicide Nisus changed into a hawk to avoid falling into the hands of his enemies. Noctua is the Latin name of Minerva's sacred owl. In contrast to *Tyto, Ninox* owls are small to large owls with relatively long tails, and some primaries are emarginated. Facial discs are indistinct, and the owls' nostrils are in front of the swollen cere (not on the sides). They are distributed from Eastern Siberia to Japan, the Philippines, New Guinea, Australia, New Zealand, Sri Lanka, Malaysia, and the Sunda and other Islands of the Indo-Pacific. There is one species on Madagascar. Australia, including Christmas Island, has six *Ninox* owls:

- Tasmanian Boobook *Ninox leucopsis*
- Southern Boobook *Ninox boobook*
- Barking Owl *Ninox connivens*
- Rufous Owl *Ninox rufa*
- Powerful Owl *Ninox strenua*
- Christmas Island Owl *Ninox natalis.*

Unlike some Australian *Tyto* species that prefer open grassland, all *Ninox* prefer forest or woodland.

Given the disagreements over taxonomy, it is difficult to say with certainty how many owl species there are in Australia. In the following chapter (in Table 3.1) we look at the taxonomy of a single owl, the Southern Boobook.

3

What is a Southern Boobook?

In this chapter we will use the Southern Boobook as an example of an Australasian hawk-owl, to underscore problems and discrepancies in the literature around a simple question: What is a Southern Boobook? Higgins (1999) classified Boobooks from mainland Australia, Tasmania and New Zealand as one species – *Ninox novaeseelandiae*, the smallest owl on mainland Australia, in the south about 300 g with the same dimensions as a Magpie. The species name, *novaeseelandiae* means the 'Hawk-Owl from New Zealand', because the type specimen is described from the New Zealand form.

But a closer look at the question: 'What is a Southern Boobook?' shows how the answer is more convoluted and contradictory than the one given above. In the Pizzey and Knight (2007) field guide and the Christidis and Boles (2008) checklist we have four *Ninox*, or hawk-owls, on the Australian mainland and in Tasmania – Southern Boobook, Barking Owl, Rufous Owl, and Powerful Owl. Offshore we have a fifth on Christmas Island, the Christmas Island Hawk-Owl. Two Boobook subspecies, the Norfolk Island and Lord Howe Island Boobooks, are extinct. Four of the five Australian *Ninox* are generally accepted but not the Southern Boobook. David Fleay (1968) in *Nightwatchmen of the Bush and Plain* discussed the difficulties around the taxonomy of this species. Mainland Boobooks, he said, look different from New Zealand Boobooks, and New Zealand Boobooks closely resemble Tasmanian Boobooks, in being smaller, darker, with golden yellow irides and pale yellow toes. In contrast, mainland Australian Boobooks have grey-greenish irides and white toes. But Fleay followed the taxonomy of Mees (1964) who concluded that Southern Boobooks from the mainland, Tasmania and New Zealand were all one species. Mees said that the classification of Boobooks was one of the most difficult taxonomic problems he had encountered.

Table 3.1. Taxonomy of Boobook species according to region and various researchers

	Australia	New Zealand (Morepork)	Tasmanian
Schodde and Mason 1980	*N. boobook*	*N. novaeseelandiae*	*N. boobook*
del Hoyo *et al.* 1999	*N. boobook*	*N. novaeseelandiae*	*N. novaeseelandiae*
Higgins 1999	*N. novaeseelandiae*	*N. novaeseelandiae*	*N. novaeseelandiae*
Pizzey and Knight 2007	*N. novaeseelandiae*	*N. novaeseelandiae*	*N. novaeseelandiae*
Christidis and Boles 2008	*N. novaeseelandiae*	*N. novaeseelandiae*	*N. novaeseelandiae*
Wink *et al.* 2008	*N. boobook*	*N. novaeseelandiae*	*N. leucopsis*

Schodde and Mason (1980) combined the Tasmanian and mainland Boobooks into one species but split them from New Zealand Boobooks, called Moreporks in New Zealand because of their call (Table 3.1). However, a standard text on birds, the *Handbook of Birds of the World* (del Hoyo *et al.* 1999), argued that Tasmanian and New Zealand Boobooks were one species, but a different species to mainland Boobooks. According to Roberson (2000) and Christidis and Boles (2008) this was a misreading of the evidence. In their checklist Christidis and Boles combined all Boobooks into one species, but Michael Wink and his colleagues in Germany disagreed. They said that mainland Australian, New Zealand and Tasmanian Boobooks were three different species, and they named them the Southern Boobook, Morepork and Tasmanian Boobook. So Wink and his colleagues recognise 12 species of owl in Australia, while the official tally in Christidis and Boles and in Pizzey and Knight listed nine species.

This is where it stands as I write. In 2010 Sue Trost and I visited Michael Wink at Heidelberg University and discussed these differences, the rationale behind his decisions about Sooty Owls, Masked Owls and the three Boobook species they nominated for Australia, Tasmania and New Zealand. They had considered measurements and other morphology in combination with cytochrome b gene analysis while other researchers had used more primitive or incomplete methods, or had misinterpreted the existing molecular data. We suspect that the Germans are correct, at least for now.

Back to the question 'What is a Boobook?' The official definition – *Ninox novaeseelandiae* should probably be *Ninox boobook* if you accept the split with the New Zealand and Tasmanian forms. Yes, it is the smallest owl on the Australian mainland, but it is larger than the New Zealand Morepork *Ninox novaeseelandiae* or the Tasmanian Boobook *Ninox leucopsis.*

Small insectivorous owl or medium-sized vertebrate killer?

Mainland Southern Boobooks are commonly described as small, insectivorous owls (del Hoyo *et al.* 1999), but those on the mainland are neither small, nor necessarily insectivorous when you compare them to other small, insectivorous owls from around the world (see Figures 3.1, 3.2 and 3.3). Using North American, South American and European owls for comparison, Australia has no small owls, only medium-sized owls and a few large ones.

In the deserts of Arizona and Mexico, for example, the tiny Elf Owl captures insects and nests in old woodpecker holes in cacti. These owls are not much bigger than a House Sparrow (see Table 3.2 below). Two other owls, the Pygmy Owl and the Flammulated Owl of the American west, are smaller than a Common Starling (see Colour Plate 2). They are less than half the weight of a Crimson Rosella, birds that Southern Boobooks often capture and eat. The Little Sumba Hawk-Owl from Indonesia, a newly discovered species and close relative of the Southern Boobook, is about a third the weight of a Southern Boobook.

According to Bergmann's Rule, individuals of a polytypic species such as the Southern Boobook tend to be larger in cooler parts of their range, like Tasmania, and smaller in warmer parts of their range, like the tropics. Canberra is towards the southern edge of the Southern Boobook's world range as accepted in current Australian texts, a range that extends from New Guinea down to the southern tip of New Zealand, and Canberra is cool because it rests on an elevated tableland. So, you would expect Canberra Boobooks to be larger than tropical Boobooks, and they are, about 300 g compared to 200 g for Boobooks around Broome or the Cape York Peninsula. But other factors affect the size and weight of birds living on islands. Boobooks (or Moreporks) in New Zealand and Tasmania live further south than Canberra Boobooks, but they are smaller and weigh less, in violation of Bergmann's Rule. However, as discussed above, the 'Southern Boobook' probably represents several species, not one.

As for the claim that Boobooks take mainly insectivorous prey, Ed McNabb (2002) argued that Boobooks in his Victorian study took many mammals, and will even attack small possums. We will discuss this in more detail in Chapters 8 and 9.

Table 3.2. Weights of some owls and two Boobook prey species

Species	Range	Wt (g)
Southern Boobook	Australasia	300
(Common Starling)	Eurasia + introduced elsewhere	75
(Crimson Rosella)	Australia	135
Little Owl	Eurasia + introduced elsewhere	170
Little Sumba Hawk-Owl	Indonesia	90
Eurasian Pygmy Owl	Eurasia	60
Flammulated Owl	North America	55
Long-whiskered Owlet	South America	47
Elf Owl	North America	41

Figure 3.1 Fledgling Pygmy Owls in Finland weigh about one-fifth as much as fledgling Southern Boobooks. They do not struggle, and they 'play dead' when they are handled.

Figure 3.2 Adult Little Owls in Spain can weigh half as much as adult Southern Boobooks.

Figure 3.3 Fledgling Southern Boobook.

So, Boobooks, around Canberra at least, are probably *Ninox boobook*, a different species from those in Tasmanian and New Zealand. They are medium-sized hawk-owls, not small insectivorous owls, and this influences the size of nest hollows they choose, the prey they eat and which animals they compete with.

Studying owls

4
David Fleay

In the first half of the 20th century much of our knowledge about breeding birds in Australia came from egg collectors, such as A.J. North, Norman Favaloro and Merv Goddard. They brought detail to mainstream science from their descriptions of bird behaviour, nests and eggs. Owls were no exception. After North, Favaloro and Goddard, a new generation of biologists pursued owls, studied them at night, photographed them and made them visible to those of us who read their books and research papers. Chief among these was David Hollands, a country doctor from Victoria. He pursued all the Australian owls, photographed them from hides at night and taught us about their home lives. His book *Birds of the Night* (1991) is a landmark in the study of owls. Before this, Dick Schodde and Ian Mason compiled most of the existing information on owls in their important volume *Nocturnal Birds of Australia* (1980). Since then others have added to our understanding of behaviour, prey, and habitat use, including Ed McNabb, Raylene Cooke, Fiona Hogan, Nick Mooney, David Milledge, Chris Pavey, Todd Soderquist, Rod Kavanagh and Stephen Debus with their research papers, and John Young with his film *Wings of Silence*. However, between the old egg collectors and this new generation of researchers was David Fleay. Serious owl research in Australia begins with David Fleay, a contemporary and colleague of those early egg collectors.

Fleay grew up in Ballarat, Victoria. His early working life started in his father's pharmacy. However, he was slightly handicapped by polio, disliked the work and his heart lay outside the town, with wildlife in the bushland where he spent many of his weekends.

His career in many ways paralleled that of Gerald Durrell, the English naturalist, zookeeper and writer. As with Durrell, Fleay ended up running zoos full of rarely seen animals. But Fleay specialised in Australian species. During his lifetime he bred a number of birds and mammals in captivity for the first time, like the Powerful Owl and Platypus. But Fleay wrote fewer books than Durrell, confined himself to Australia and took less advantage of television to publicise his achievements. He is less well known, but he probably contributed more to our knowledge of animals than did Gerald Durrell.

Owls were his favourites, and he described Southern Boobooks as the 'Last of the owls to nest each season, the rich brown Boobook usually has its fluffy, comical owlets on the wing in early December.' He went on in *Nightwatchmen of Bush and Plains* to say, 'As a child of nine or ten, I fell hard for Boobook owlets … The sight of three tiny white ghosts, softly downed, trilling like crickets, and winding their heads round and round in the liveliest manner at the wonder of the night world, enchanted me as nothing ever had before.'

In 1924, as a schoolboy, he built an aviary enclosing a live apple tree, a dead pine and part of a Silver Wattle for some Boobook chicks he took from a nest. He intended to watch their behaviour close up as they developed. He learned early how they needed whole animal prey, mice and sparrows, to stay healthy. He described an owl preparing to disgorge the indigestible parts of fur, bone and feathers in a pellet: 'His eyes take on an expression which is distinctly oriental; he yawns repeatedly and finally loses the small object containing perhaps finely broken bones from the sparrow that was his last meal. Sometimes the pellet is disgorged into the receptive grip of talons and drowsily inspected before final rejection.'

His captives skilfully caught dead mice tossed to them in mid-air, and they snared live starlings or sparrows that squeezed through the wire netting from outside, probably looking for food. The owls took them to a perch, and killed them with a bite to the junction of the skull and neck, as falcons do. He described changes in the owls' plumage through the years, and their voice which, to Aboriginal people, sounded like 'buc-buc' or 'boobook'. This supplied the vernacular. Early English colonists thought they were cuckoos calling at night.

Fleay was the first to breed Boobooks in captivity, describe incubation, growth of nestlings and how the young fledged (flew) at about six weeks old, 'winding their pretty heads round and about at everything, and especially the flickering shadows cast by the hurricane lamp I carried'. Much of what we know about Southern Boobooks is built on Fleay's early work.

Powerful Owls

The Powerful Owl *Ninox strenua* is the 'strong' Hawk-Owl (see Figure 4.1). As with Southern Boobooks, the genus name, *Ninox*, comes from two words: Nisus, in

Figure 4.1 Powerful Owl.

Greek mythology the King of Megara, and Noctua, the Latin name of Minerva's sacred owl. The species name, *strenua,* comes from the Latin *strenuus* – 'vigorous' because of its very large size.

The wild haunts of the Powerful Owl make them hard to find and study. Still, they are favourites of Australian ornithologists going back to the 1920s. David Fleay, and some might argue, Merv Goddard, are the fathers of Powerful Owl research. In *Nightwatchmen of Bush and Plain*, Fleay described how, as a young man, he began to search for Powerful Owl haunts in the high Victorian forests. He would cycle for 20 miles over rough logging tracks to meet the elderly Smith brothers at the tangle of tubes, brass pipes and galvanised iron roofing they used to distil eucalypt oil deep in the peppermint forest. With the Smith brothers, who he said laid a scent of eucalypt wherever they went, Fleay hiked at night and watched Greater Gliders, a metre long, saunter to the edge of a branch, gather themselves, and leap, with their long tail streaming behind them, to alight on the trunk of a target tree. And they followed Powerful Owls that followed the gliders, became lost more than once and had to sleep on the ground until dawn, then had to find their way out.

He tells of finding a Powerful Owlet in 1926 he called Ferox, abandoned at the Ballarat Zoo, sitting forlorn and sick over slimy green water in a clapped-out Water Rat pen. He raised it and took it with him to the Toorak Central School in 1931 where he began teaching. Later, in 1941 and 1942 he began studies of nesting pairs of Powerful Owls in the Victorian forests.

Up to 1945 there was only one set of Powerful Owl eggs known to science, described in the journal *Emu* in 1912. They were said to have been taken on 23 November 1911 at Kingaroy, Queensland, and kept in the H.L. White Collection in the National Museum in Victoria. Fleay thought this clutch was a fake, partially because the eggs were taken in spring, and Powerful Owls breed in autumn, and partially because a charlatan named A.E.L. Bertling had sold a number of suspect clutches, including the alleged eggs of the Fig Parrot, to egg collectors. It was Bertling who had sold this particular set of Powerful Owl eggs. Oologists, as egg collectors preferred to call themselves, were prominent in Australia in the early 1900s and some, especially graziers, had enough money to attract swindlers. The eggs in the museum were smooth, glossy and slightly pointed at one end, not rounded as other *Ninox* eggs. Fleay thought these eggs were more likely the eggs of a Eurasian Eagle Owl or Great Horned Owl, deliberately faked to make a sale. Eagle owls can nest on rock ledges and open stick nests made by other birds, which Powerful Owls do not do. It would make sense that eagle owl eggs, unlike Powerful Owl eggs, could be slightly pointed at one end to avoid rolling off cliff ledges, though Great Horned Owl eggs I have seen were similar in shape to Powerful Owl eggs but slightly larger.

So Fleay believed there were no Powerful Owl eggs in collections, and no description in the ornithological literature. He set out to change this. On 4 June 1945, after a long search, he took the first set of Powerful Owl eggs for science, from 'rough and dense habitat' between Mount Toolebewong and Mount Riddell near Healesville, Victoria. He described the eggs as smaller than those in the museum, more rounded, dull white with only a suggestion of gloss. As the eggs were taken early in the breeding season the parents re-nested and raised two young.

At the time, his friend, Merv Goddard, a famous amateur ornithologist from Tenterfield in northern New South Wales, was as obsessed about Powerful Owls as Fleay was, especially about their eggs. In those early days Fleay and Goddard would camp out in tall forest together and follow the owls through the night. Goddard was a superb climber, well into his forties, and he climbed to a nest, with Fleay watching from below, and took two nestlings for Fleay to foster under his broody but infertile pair of captive Powerful Owls.

Since Fleay first saw Ferox as a chick in the Ballarat Zoo, he was determined to breed the species in captivity, as he had other owls and raptors, the Platypus and more. The task proved near impossible. Captive Powerful Owls would die, some from an air-sac mite, some would fail to lay eggs, or the females would incubate infertile eggs until they addled. In his book he says with a sense of victory that on 13 July 1968, 'the culmination of 42 years dogged by unnumbered misfortunes', he bred the first Powerful Owls in captivity.

In the late 1970s, an old man now at his zoo in Burleigh Heads, Queensland, he seemed tired, worn out by all that he had done and he did not give us much of his time. But he smiled and a glint came into his eye when he talked about Powerful Owls.

Merv Goddard had always been withdrawn, but, by the late 1970s, he had become deeply suspicious, a loner who refused to see most visitors. He did not allow anyone to see his egg collection, and allowed a compelling rumour to flourish – that he had gone a little loopy and smashed the collection, knowing that it was supposed to be the most valuable egg collection in Australia, and that he was, arguably, the most knowledgeable ornithologist in Australia. A few years earlier, the curator of birds at one of the state museums had gleaned all he could from the detailed cards accompanying Merv Goddard's egg collection, cramped notes about eggs and nests and the behaviour of birds – and had then turned Goddard in to the authorities for illegally collecting eggs.

They never found the Goddard collection. Maybe he had smashed it, or secreted it in the attic of a sister, or son or maybe a friend. But that was the end of Goddard's cooperation and fellowship with professional ornithologists. He was finished with museums. David Fleay immortalised him anyway in the book that started owl research in Australia.

5
Surveying owls

It is harder to study owls in Australia than in North America or Europe. This contrasting difficulty shapes much of the research done, and the type of information published. We know less about the behaviour of Australian owls, like the Powerful, Sooty and Rufous Owl than we know about large or medium-sized owls in Europe and North America such as the Great Horned, Tawny and Great Grey Owl, and there are good reasons for this. Paradoxically, Australia still has many large trees and owls have a choice of nest sites. They seldom use nest boxes. In Europe, where forests were long ago felled, entire populations of owls breed in nest boxes nailed or wired to trees, boxes that are easy to access and study. Researchers do not search for nests because they put them there in the first place. Also, many European owls nest in buildings. Owl nests in Australia are much more difficult to find.

Visibility is another problem. Driving along country roads in North America, even in summer, you can see Great Horned Owls from the window of your car, sitting in poplar trees along the roadside or hunting at dusk. You can see daylight-hunting Short-eared Owls, Pygmy Owls, Northern Hawk-Owls and other species. In winter around Winnipeg, Canada, you can see owls hunting at noon. No owls build a nest, and North American owls such as Great Grey Owls or Great Horned Owls often take over the stick nest of a hawk, or raven, a nest clearly visible from below. In places like the dry eastern border of Washington State, you can find Great Horned Owls in the abandoned nests of Red-tailed Hawks in Ponderosa Pines. Brooding or incubating females are conspicuous; they often stand next to their nest in bright sunlight, their broad faces and feathered 'horns' resemble a staring cat, eyelids part-closed as if unconcerned. In Australia we have no such luxury. Australian owls incubate deep in hollows or caves, often in forest. Australian owls almost never hunt during the day, and almost never use stick nests. Most come out only at night. They are invisible.

Tall, shaded forest is typical Rufous Owl, Powerful Owl, Sooty Owl and Masked Owl country. Instead of seeing an Australian owl, you often hear them call, and then only at certain times of the year. This is important because much of the study of Australian owls is done by inference and deduction. Researchers seldom watch owls hunt and behave. Key principles of owl behaviour – how they use a patch of land, defend territories, forage or interact with enemies – are little known. Researchers often estimate population numbers or classify an owl species as Endangered, Threatened or Vulnerable by listening to owls calling at night or collecting prey remains. For example, in Newton *et al.* (2002), 13 of the 17 papers on Australian owls were either diet studies and/or analysis of playback protocols. Only two were observational studies. Relatively few nests are found, and fewer nests are climbed, even with common species like the Southern Boobook. Researchers go into the forest at night, stand next to a vehicle, play a recording of an owl, listen for the owls to respond, then climb back into the vehicle, drive to another spot, play the recording again, listen for owls again, then deduce as much as possible from the frequency and nature of the responding owls. A standard five-step protocol could involve:

1) identifying survey localities
2) at each locality, listen for 15 minutes for unsolicited vocalisations
3) play a pre-recorded broadcast of the owl species for 2 minutes and wait for a response
4) spend 10 minutes searching the plot with a 100-W spotlight, and
5) mark on the map where owls called or did not call (present or absent).

Looking at an example in more detail, McIntyre and Henry (in Newton *et al.* 2002) used the following protocols to survey large forest owls in East Gippsland, Victoria, to generate a model of preferred owl habitat when they conducted a census of 440 sites between 1991 and 1993. They stratified these census locations to include all Ecological Vegetation Classes – coastal vegetation, heathland, wetland, banksia woodland, lowland forest, riparian forest, shrubby dry forest, grassy dry forest, box-ironbark forest, limestone grassy forest, woodland, rocky outcrop scrub, damp forest, wet forest, rainforest and montane forest – covering more than 1000 ha in the region, approximately in proportion to their extent. They also used four age-classes of forest – senescent dominant, mature dominant, mixed age classes and regrowth – and indices of steepness, distance to the nearest patch of heathland, rainforest or old-growth forest. They carried out their observations during the first four hours of night and investigated each census site once, throughout the year, mostly between November and March. They broadcast taped calls of Powerful, Sooty and Masked

Owls using a Walkman and megaphone. This was followed by an active spotlight search and listening period of 30 minutes in the vicinity of the census site. Each census was deemed to sample a circle of 1000 m radius, equal to 315 ha, the distance over which a human could readily detect the broadcast calls on a still night. The estimated location of each responding owl was mapped.

They used the results of this survey, the locations where they detected or did not detect owls calling, to protect habitat for 120 pairs of Powerful Owls (old trees especially in wet, damp and riparian forest containing arboreal mammal prey), 131 pairs of Sooty Owls (old trees in lowland damp and riparian and warm temperate forest with large hollows for breeding and small terrestrial and arboreal mammal prey) and 103 pairs of Masked Owls (older age classes of lowland forest and coastal woodland, and riparian forest on flat terrain, heathland and banksia woodland, with small terrestrial mammal prey). Researchers did not locate the nests of breeding pairs. This would have been too difficult on such a scale. They inferred appropriate habitat for breeding pairs from playback surveys, and diet studies done elsewhere.

Liddelow *et al.* (2002) used similar survey techniques on public forest land in Western Australia during the spring of 1999 and autumn 2000. Concentrating on Masked Owls and Barking Owls they surveyed 70 sites, stratified by major forest type (Jarrah, Karri, Wandoo) and logging history (unlogged, logged once or twice, logged three times or more). Using taped playback protocols similar to those described above, once in each season, they located a total of 196 Southern Boobooks and 15 Masked Owls at 67 per cent of the sites. Within the main forest belt 156 owls were recorded with 55 located in the outlying forest sites (see Figure 5.1). No Barking Owls were recorded.

Some might question this finding. In the winter of 2007 a single Powerful Owl turned up in Canberra at the National Botanic Gardens on Black Mountain. Rangers Greg Flowers, Joe McAuliffe and Greg Sattler recorded its call on a cell phone and played it back. It responded once, but not again. We broadcast another Powerful Owl call close to this owl and, again, it did not respond. Boobooks on Black Mountain did the same, even though we had not previously used playback in the study. Radio-tagged owls often did not respond to the broadcast of a Boobook call, or they responded during the first few nights of playback, then went quiet. In a survey of broadcast calls it would look as though Boobooks or Powerful Owls had disappeared from Black Mountain. However, they had simply gone quiet. Maybe they did not see the playback as a threat, or had been exposed to a lot of playback – and this is one limitation of owl playback surveys.

Because there is so little known about owls in south-western Western Australia – researchers have located few nests, done no major studies of behaviour, diet or hunting – this form of survey will dictate much of what is reported in the current literature, especially about the Barking Owls Liddelow *et al.* failed to locate. Given the difficulties of studying owls in Australia, this form of data is a mainstay.

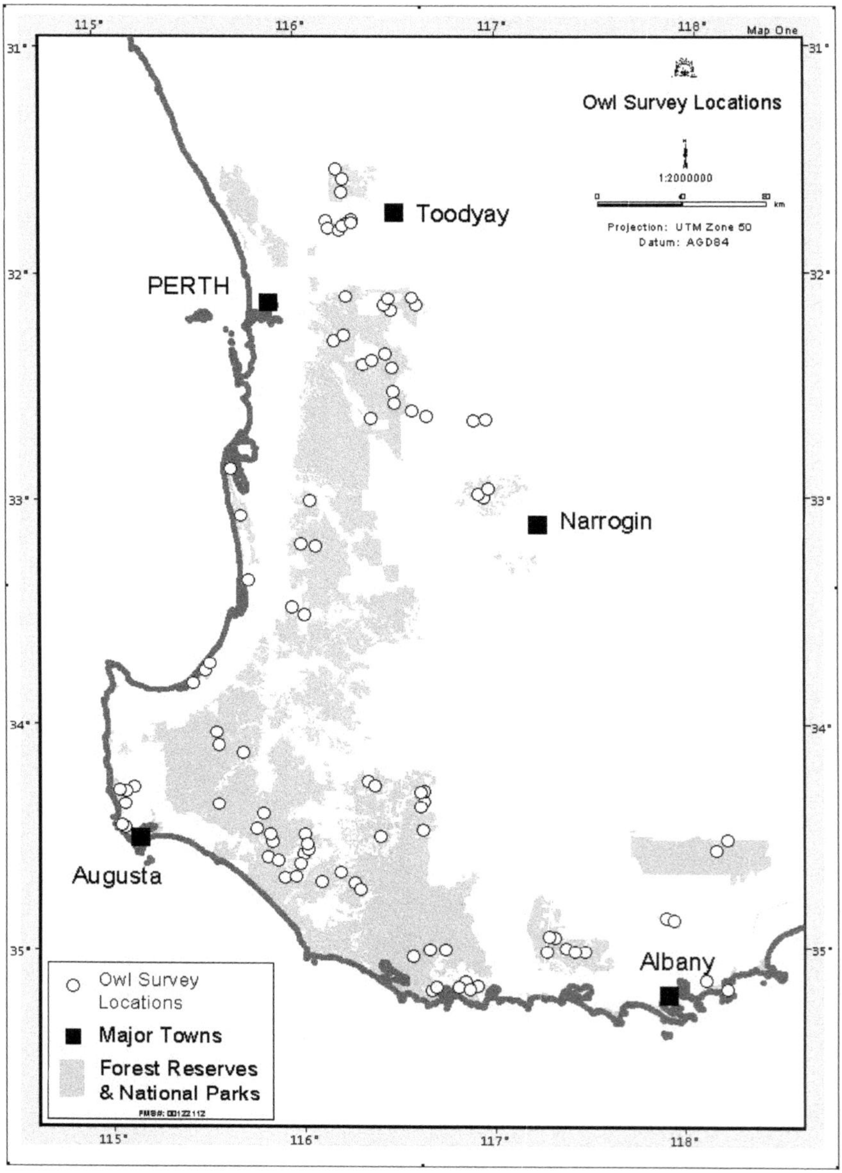

Figure 5.1 Distribution of survey locations for Masked Owls and Barking Owls from the Western Australian Museum, and from information from the general public on Barking Owls (Liddelow *et al.* 2002).

The above are important studies. This is not an attack on their methodology. The point is that our portrait of certain owl species in certain areas of Australia is limited to survey data. We should take care with conclusions drawn solely from surveys of difficult-to-study owls.

6

Finding owls

Finding owl roosts

As mentioned in Chapter 5, few surveys of owl vocal responses locate the owls' nests or roosts. For example, Kavanagh (2003) discussed the conservation of owls in Sydney's urban bushland but found no Boobook nests, though Boobooks were considered common and often heard. Actually locating owls and their nests is different from locating vocalisations on a map.

Owls can be located by day. *Ninox* owls often prefer bushy trees for day roosts. David Fleay said that wild Boobooks around Ballarat often roosted by day in the branches of Native Cherry (*Exocarpos*) trees. Australian owls are cryptic and still by day. If they move, other birds will often attack them – this can be a good way to find owls close to dark when they start to become restless and stretch and move. Currawongs, friarbirds, Noisy Miners, Common Mynas, wattlebirds, fantails and other passerines give alarm calls and can reveal owls concealed in vegetation (see Figure 6.1).

Owls often moult during the breeding season and primary or tail feathers drop under roosts. During midday in spring or summer, researchers sometimes find a delicately curved breast feather or barred primary on the forest floor, or a white splash under a eucalypt, signs that owls have been there. Researchers check the owls' favourite roost trees and grass or eucalypt foliage beneath with binoculars. Powerful Owls leave pieces of glider such as tails or entrails looped in the canopy or on the forest floor.

Figure 6.1 Birds such as Noisy Miners give alarm calls when owls are emerging from their nest hollows.

Boobooks, when found, stand upright and alert; they watch you as you watch them. Sometimes they roost at eye level in Native Cherries, and walkers and joggers pass by within metres and do not notice. But the owls can become more alert, sometimes nervous, when they perceive that a human is interested in them, watching them and not just passing by, especially if the observer has binoculars or is pointing a camera. Owls respond to us as predators, which, of course, we are, not intent on eating or harming them, but they don't know this. Those who understand owls do not approach them too closely in low trees like Native Cherries because the owls can flush in daylight and currawongs or other birds can then attack them. Experienced adults can protect themselves from the beaks of enemy birds, but fledglings are comparatively defenceless in daylight, and they can be hurt or even lose an eye.

A family of Powerful Owls or Southern Boobooks roosting in a tall eucalypt tree tilt their faces downwards at 90 degrees, watching the observers. Sometimes they sit in a row along one limb, shoulders touching, brown-feathered adults next to cream-coloured juveniles (see Figure 6.2). In the first few weeks after fledging, females may sit alone with the young; adult males can sit up to 50 m away. Researchers collecting prey remains might gather these underneath the family group, then search nearby for the missing adult all the while collecting below him or her. Just-fledged Boobook young are stubby-tailed with round bodies and dark bandit masks, like raccoons. They fly clumsily at night and give persistent cricket-like trills.

Figure 6.2 Fledged Southern Boobooks and other owls often roost together by day.

Finding nests

Researchers sometimes locate fledged owls in spring or late summer, then return the following year and search for a nest. Because juvenile Boobooks are so noisy around Christmas and January, and Powerful Owls noisy around September to November this can be the best time of year to locate territories, but not necessarily the best time of year to locate nests. The family can move some distance from the nest, camp out each day in different parts of the territory, or change nest sites in the following year.

There are profitable as well as unprofitable times to listen for calls – August to October are good months for calling Boobook adults and December and January for fledged trilling young. March to May are good months for listening for calling adult Powerful Owls, September onwards for trilling young. Researchers find owl nests by listening more than watching, but not necessarily by listening for territorial calls.

By winter, juvenile Boobooks have dispersed. At dusk in early spring, the adult owls tend to drift away from the nest site without making a sound. The forest is peppered with likely nest hollows – in tree trunks, empty dead limbs, broken-off spouts, holes just big enough to allow the owls in but keep out a cat or Common Brushtail Possum, about the size that would allow passage of a champagne bottle. Higgins (1999) said that Boobooks call from vantage perches mostly near the nest site. This should make nests easy to find, but if you look at various surveys of owls you find that researchers map the locations of Boobooks by their calling, but almost never find their nests, even though they may know of Powerful Owl nests or Tasmanian Masked Owl nests in the same area. Few Boobook nests are known for Tasmania, where they may be a separate species. Boobook nests are hard to find during incubation because adults are secretive.

Though suburbs have dozens of suitable trees, Boobooks usually nest in forest away from suburbs. Suburbs and the city are too chaotic, with too many cats, dogs, people and possums. Because owls take such care to conceal their nest hollows, much of the task of finding a nest relies on listening to the *type* of call. People often report a calling Boobook or other owl, and assume the nest is close by. This is seldom the case with Powerful Owls or Boobooks. Mostly they use far-carrying territorial calls some distance from their nests.

This was tested with two male and two female Boobooks between 1996 and 1999. Their nests were located, and the distance from the nest that each owl called using 'Territorial Boobook' calls, the normal call associated with Boobooks, was measured. The distance from the nest and number of 'Bray' calls they used were also scored. The Bray call sounds something like a Wood Duck or a baby magpie calling for food and were used only by females, to solicit food or assistance from males.

The frequency that males or females gave 'Boobook' calls inside a radius of 100 m of the nest, or that females gave 'Bray' calls inside this radius was scored. The two females used 'Territorial Boobook' calls within a 100-m radius of the nest tree only 12 per cent of the time, the two males only 31 per cent and 0 per cent of the time (Table 6.1). The median distances from the nest that two females gave 'Territorial Boobook' calls was 175 and 250 m, for the males 210 and 525 m from the nest (Table 6.2). That is, the locations of 'Territorial Boobook' calls on a map were not very helpful clues for locating nests because owls often called on territory borders.

Median distances from the nest that the two females called using 'Bray' (food-begging) calls were 10 m for both females; they called within 100 m of the nest 70 per cent of the time for one female and 88 per cent for the other (Table 6.1).

Table 6.1 Frequency of bouts of 'Territorial Boobook' and 'Bray' calling by females F White, F Green and males M Green, M White, <100 m versus >100 m from the nest being used at that time

	Boobook calls		Bray calls	
Subject	<100 m	>100 m	<100 m	>100 m
F White	3	23	39	17
F Green	2	15	80	11
M Green	17	38		
M White	0	15		

Table 6.2 Median distance from the nest in metres that two male (M Green, M White) and two female (F White, F Green) Southern Boobooks called using 'Territorial Boobook' calls (n= number of bouts of calls recorded for that individual)

Subject	Year	n	Median (m)	Range (m)
F White	1998–99	26	250	10–950
F Green	1997–99	17	175	0–450
M Green	1997–98	56	210	2–600
M White	1996–98	15	525	200–1225

Though 'Territorial Boobook' calling was not an effective way of finding Boobook nests, listening for 'Bray' calls of females did help us find nests. This seems to apply to larger *Ninox* owls as well, such as Powerful Owls which have an equivalent two-note territorial call, a deeper *who-who*, and an equivalent 'Bleating' food-begging call used by females, though David Hollands (2008) said that both sexes use this 'Bleating' call near the nest. As with Boobooks, these latter calls are more useful for finding Powerful Owl nests.

Two other calls help researchers locate Powerful Owl and Boobook nests. The Trilling of young inside the nest hollow can help, although these calls are difficult to hear and of no use early in the breeding season. The muffled 'Contact' calls that males and females utter close to each other, like the two-noted territorial call but quieter, are helpful clues but cannot be heard at a distance.

Finding owl nests in Australia remains more of an art than a science.

7

Trapping and handling owls

Other methods of studying owls include telemetry and colour-marking adults. First, you need to trap an owl, and this can be difficult. Some researchers broadcast taped playback of their calls to lure owls into standing nets. This does not work very well with Southern Boobooks, but it does work with Moreporks in New Zealand, and Powerful Owls. See Stephenson (1998) and Soderquist and Gibbons (2007) for details on how to use this method. Below are some examples of other trapping methods commonly used.

Bal-cha-tri and 'flip-traps'

We carried a 'flip-trap' into the field, a sort of automatic bow net (see Colour Plate 8) to catch an adult Boobook feeding a single fledged young. The trap contained a live mouse. Three principles are at work here: (1) owls catch prey with their feet, not their beaks, (2) an owl will see the wire mesh protecting the mouse as dense cover they can crash through, (3) nothing gets hurt, including the mouse or the sparrow used to lure a raptor into a trap.

If an owl attacks the mouse, and hits the trigger-plate on top, a spring-loaded net swings over and entangles the owl. We rush in and grab the owl before it untangles itself and escapes.

Another trap, the bal-cha-tri, a traditional trap used in India, uses similar principles. Nooses on top snare a raptor's feet if the raptor attacks the mouse. The owl 'thinks' it can get through the wire mesh and carry off the mouse, but the trap has two layers of mesh on top, spaced about a centimetre apart, that protect the mouse from the raptor's sharp talons.

We located an adult owl after dark, then looked in the torchlight for open flat ground between tussocks, away from the metre-wide domes of bare soil in the field. The domes are ant nests spaced over the ground, and ants can attack the mouse.

We left the trap where the owl could see it, away from ant domes, and waited. Lab mice aren't frightened inside a trap. They sniff the grass and poke their noses through wire mesh. Most are stimulated by trips out and away from their plastic homes, paper nests and running wheels. Their restless exploring attracts the owl. Traps are only effective if mice roam and sniff around. If the mouse huddles in a corner, the owls do not see it and the trap does not work. The best trapping mice are old pets comfortable in traps.

We stood 50 m away with binoculars. The adult sailed over the field, flipped on its back, caught an insect, glided to a branch next to a trilling young. He turned his head (we later determined it was the male parent) and carefully placed the insect into the fledgling's mouth. The trap was 20 m away, but he ignored it.

He flew sortie after sortie catching insects over the field, then sat next to the trilling juvenile. He looked down at the trap; the juvenile ignored it. The fidgeting mouse sniffed and explored grass and twigs through the wire mesh. The adult dropped its head low and stared for 20 seconds, then glided down from the branch, hit the wire-mesh plate on the roof, and triggered the bow net. The net swung over in a blur.

We ran in, grabbed the struggling owl, untangled it from the net. The fledgling sat and watched.

With the newly trapped owl we followed a routine – measure its wing, tail, one leg and a middle toe with callipers and a ruler, squeeze on a metal band and write the number on a data sheet next to the band numbers, then glue a plastic red band on the owl's other leg. Only certain colours work on nocturnal birds because a lot of dark blues and dark greens look black at night – white, yellow, red and light green bands work best.

Ten minutes after trapping, we turned it loose.

Noose pole

Tracy Fleming's Spotted Owl study areas were in the eastern Cascade Mountains of Washington State near Glenwood and Mount Adams. You travel along the state's underbelly skirting the Columbia River, as Lewis and Clark did in 1812. Washington is on the north bank, Oregon on the south. The river is 2.5 km wide and the road winds under and up over the tops of cliffs, rolling down into quiet mountain villages. Pick-up trucks in the main street of Glenwood had rifles saddled in their rear windows.

Figure 7.1 Spotted Owls are cryptic against tree trunks.

Following Tracy into shaded forest at 9 a.m. the walking was easy, as comfortable as walking in a city park. Most trees, Tracy said, were some 90 years old. He cupped his hand around his mouth and gave the four-note call of the Spotted Owl, and listened. After a minute, an owl answered and he walked towards the hooting. In a dark glen sat a roosting adult, cryptic against the trunk of a Douglas Fir (see Figure 7.1).

Tracy brought out a white lab mouse and let it run along a leaning branch under the owl. The owl watched for a moment then plunged off the limb, caught the mouse, killed it and swallowed it whole. The owl flew up to a lower limb, and Tracy placed another mouse on a limb to induce it down further. The owl dropped off the limb, caught the mouse and devoured it.

Tracy pulled out a long collapsible pole used to catch panfish or bream, sometimes called a Bluegill fishing rod. Secured on its tip was a salmon leader, wire coated in plastic, and made into a noose about as round as a teacup saucer. After he released the third mouse, the owl perched lower. Tracy extended the pole up towards the owl's head. The noose had a tab of gaffer's tape placed to stop it from closing, so the noose would go over the owl's head and capture it by the neck, but not strangle it. When tightened, the noose would form a collar stopped by a tab, and the collar would keep the owl from flying away.

Figure 7.2 Spotted Owl in Washington State captured using a noose pole.

Tracy moved closer and dropped the noose over the complying owl's neck, then pulled it tight. The owl tried to fly off but fluttered to the ground where Tracy and his assistant secured it. They weighed, measured and colour-banded it. The owl struggled, probably expecting to be eaten, but then settled, rested calmly and looked at us (see Figure 7.2). Tracy released it and watched it fly to a low branch in a Douglas Fir. It ruffled its feathers and preened.

It is illegal to use mice in this way in Australia, but the noose pole is a safe way to catch roosting Australian owls, especially Boobooks, if the mouse is not used. Providing the noose doesn't tangle in branches, and the researcher can climb to the trapped owl if it does tangle, the owl will not be injured.

Noose pole and moving hide

The steep forested hills on Chuckanut Drive in Washington State open to flat ground near Edison and Bow, the Skagit Flats farming villages with rain-soaked pastures. Every winter these flats draw hundreds of raptors – Bald Eagles, Gyrfalcons, Peregrines, Merlins, Rough-legged Hawks and, in the winter of 1996–1997, hundreds of white Snowy Owls. Many sat hunched in the rain on fence posts or telephone poles, or slit-eyed on rooftops in the sun. Others sat on the ground among machinery or on top of discarded piles of farmyard lumber.

Snowy Owls live in the Arctic, but in 1997 many irrupted southwards. On the left were the snow-covered Cascade Ranges, and on the right was standing water in ditches and small lakes in the middle of paddocks. A flock of Trumpeter Swans rose calling, and six or seven Short-eared Owls sat on open ground away from Snowy Owls.

Bud Anderson and his team were trapping Snowy Owls to wing-tag as many as possible before the owls returned to their Arctic breeding grounds. Wing-tags are hand-sized flaps of leather or plastic fabric that biologists fold over the leading edge of a bird's wing and fasten with rivets.

Researchers pierce a hole in the patagium, the thin membrane of skin stretching from the body of the bird to the carpal joint of the wing, and fasten the tag with a rivet through the hole.

The effect of these colour-bands, when the owl sits broadside on the ground, is a white bird with a brightly coloured, numbered sign lying flat over its folded wing. If another biologist spots this owl on its Arctic breeding grounds, they can report the number to a central banding authority, and banders share its location.

Outside, a chilling fog settled over pastureland. Two Snowy Owls had settled on the peaked roof of a two-storey farmhouse, an odd sight. Both were striking, snowy-white birds the size of small turkeys with black hatching on their backs and bellies, their shiny black beaks part-concealed by feathering over their nostrils. Their almond-shaped yellow eyes were bright against blue winter sky.

Bud asked the owner if he could trap the owls on his roof. The man pointed skywards, seemed to know about them, and nodded 'Yes'. Bud set up a long pole with a fishing line noose at one end, and threaded the line down along the pole to his free hand. He planned to loop the owl's feet and pull the noose tight. Slowly he moved the pole towards one of the owls, and dragged the noose up the sloping roof. Progress was painfully slow. The owl watched him but didn't seem overtly perturbed, but when the pole inched closer to its feet, both owls flew.

Other attempts failed. Bud decided to try owls perched lower, on fence posts in fields. He had devised a walking hide, a 2-m high cylinder of wheat sheaves cut big enough to hold a person, and he used this portable hide to edge slowly and cautiously to within pole range of a Snowy Owl roosting on the ground, or on a fence post (see Figure 7.3). He approached the owl, taking half an hour to creep within touching distance with the pole. The owl did not fly. Even when he accidentally tapped one on the shoulder with the pole tip, the owl reached down and brushed the spot with the side of its head, as if relieving an itch. Then it flew.

Looping their feet was going to be difficult so Bud decided to loop them around the neck, as researchers did with Spotted Owls in the south of the state.

Figure 7.3 Bud Anderson inside a moveable hide, moving towards a Snowy Owl on the Skagit Flats.

The noose-pole would be particularly safe on the grassy Skagit Flats because it could never tangle in branches.

In the first attempt, the owl was perched half a metre off the ground between farm machinery along a part-frozen road skirting bright green pasture. Bud moved a few painful centimetres at a time in the cylinder of straw, like a shuffling Geisha over the frozen mud. Ice crackled in puddles and the owl stared, as if curious.

Bud stopped for a minute, waiting in cold sunshine for the owl to look away, the end of the pole quivering in the breeze, like a medieval knight with a long, drooping jousting pole. He had been stalking this owl for nearly 30 minutes and must have been shivering in his porous costume. He moved another 2 cm. More ice crackled. He waited. Slowly the tip of the pole rose towards the owl's head; the owl stayed put and he looped the noose a millimetre at a time over the owl's white head. With a sideways pull he closed the noose and the owl flapped away from the farm machinery. It pulled but was caught, staring at Bud and beak-clacking (see Figure 7.4).

Handling trapped owls

Normal procedure for picking up an owl is to place a hand in the middle of its back and press down so that its legs extend backwards out under its tail. Then encircle the body, including the wings and legs, by wrapping both hands around the owl's

Figure 7.4 Trapped Snowy Owl with patagium wing tag.

lower back just above where tail feathers emerge. Pick it up with its beak facing away from you and carefully disengage claws from the netting or foliage. Place a finger between the owl's tarsi so it does not grab its own feet (as shown in Figure 7.5).

(Raptors are powerful animals. When held by humans, they can struggle and break feathers, hurt themselves or hurt the handler with their powerful beak or claws. A biologist working long enough with owls, hawks, falcons or eagles will at some point be grabbed by the raptor he or she is trying to measure and band – 'process' as biologists call it. The raptor slashes out with needle-sharp talons at any part of the biologist that strays into range, and the grip is so powerful that talons can be hard to detach.)

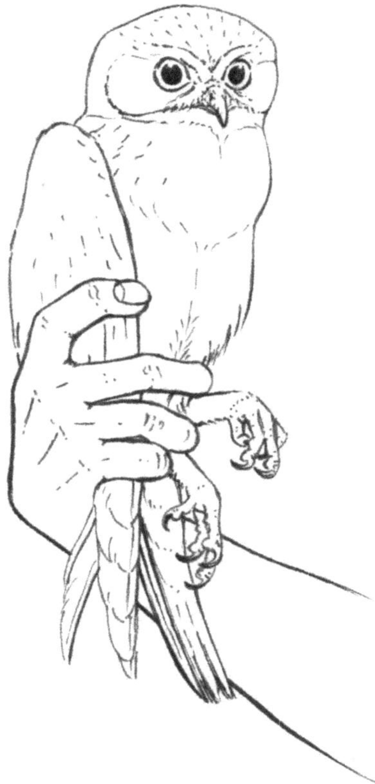

Figure 7.5 Holding a Southern Boobook.

Remain vigilant while holding the owl. *(Owls seem to have short legs, but this is an illusion. They have long legs, long tibia and tarsi. The legs look short because their body and flank feathers hang down over the top half of their tarsi. Southern Boobooks reveal their long legs when they hunt. They extend their legs with their feet wide open and hit the tree canopy, or chase birds into the tops of bushes and clutch them with their outstretched legs and toes.)*

While being held, a Boobook's long legs can shoot out, grab you and puncture you with four talons. If this happens, follow a few simple rules: If alone, hold the owl firmly so it cannot free the other leg and claw you with the other foot. Do not make jerky movements that can hurt the owl, or dig the talons in deeper. You, or an assistant, need to straighten the owl's leg, but not by pulling the leg out. Instead, grasp the femur, tibia and tarsus and gently force the bends out of the owl's leg so the locking tendons that hold the four grasping toes curled inward (and curled into you) are released. *(This tendon arrangement gives the owl its powerful grip, and allows birds to sleep on a perch without falling off. Understanding this mechanism, you can use it to your benefit. With the owl's leg straightened, remember that talons are deeply curved, not straight, and each talon is concave underneath, not tubular. That is, if you pull the talons straight out of you, they will pull out your flesh with them.)* Carefully pull out each talon around the curve of its shape, one at a time. Then disengage each talon from your clothing.

~

On the Skagit Flats, Dave, a student, held the big Snowy Owl carefully. To attach the patagial tag, he pierced a hole in the patagium, folded the tag over the leading edge of the wing and fastened the tag with a rivet through the hole. But the owl struggled and managed to fasten all four talons into the palm of Dave's hand. This is a much more serious matter than being grabbed by a Boobook. Snowy Owls are large, powerful predators. Dave squatted and grimaced; the pain unbearable. We straightened the owl's leg, releasing the tendons holding the toe-locking mechanism. The Snowy Owl could not help but spread its toes. The stabbing pain took the colour out of Dave's face. He felt a rush of nausea, a feeling of heaviness and the need to lie down.

After a few minutes, Dave recovered, finished the measurements and released the owl.

8
Telemetry

Owls can spend most of their time in a small area of their home range and rarely visit other parts. To track owls and determine point locations inside their home ranges at different times of the night, researchers attach small battery-powered transmitters that emit low-powered signals via a small antenna. Through radio-tracking researchers can also determine home range shape and size, and how often owls use certain parts of their home range compared to others. Researchers glue small acrylic-covered transmitters to the central tail feathers, or tie them to the back of an owl with a string harness that fits like a vest – a wing goes through a right and left loop, and the radio is tied and glued behind the neck and midway between where the wings emerge. This point between the wings is the centre of gravity for an owl and a radio attached there will not impede flight. Built into the harness is a weak link that sits over the breastbone; made of a single thread, it is designed to break if the transmitter or harness tangles in foliage (see Figure 8.1). The radio detaches if it snags on bushes or thorns, and this is important because Boobooks hunt birds and insects in bushes, and in the forest canopy.

Radios used with Boobooks weigh about 5.4 g and the harnesses about 1.0 g, making 6.4 g of added weight, 2.4 per cent of the body weight of a 270 g male Boobook, and 1.9 per cent of the body weight of a 340 g female Boobook. Batteries last 10 to 12 months. These radios give a regular beep heard through a receiver and earphones attached to a directional hand-held antenna. The beep is louder when the antenna is pointed towards the owl, and weaker or silent when it is pointed away. The signal can bounce in hilly or wooded country and this reflection or refraction of the signal can create false bearings, so you need to walk or drive to another location and take a second or third reading to get several readings and determine the owl's precise location. This is called 'triangulation'.

Researchers disagree about the meaning of these point locations on maps, but most raptor biologists use a minimum convex polygon (MCP) to show the size and shape of the home range. This method draws a line around the outer edges of the plotted points and assumes that the outer edges of the home range are convex, not concave. This estimates the 100 per cent MCP but many biologists say a 100 per cent MCP over-represents the size of the home range. They construct a home range from 95 per cent of the points, and a 'core area' estimate of 50 per cent of the densest points. This helps to remove the misleading effects of random outlying points. The notion is that raptors stray occasionally to these outer points but seldom use them, so the points are not part of the normal home range, they do not represent the true outer edges of the home range. Some researchers, for example Downs and Horner (2008), also criticise MCP as a measure because true home ranges may not have convex edges; they might have concave edges. But MCP is commonly used with Australian owls (Soderquist and Gibbons 2007; Kavanagh and Jackson 1997; Kavanagh and Murray 1996) – and allows for comparison with other studies. The following example uses MCP.

In Canberra, many adult female Boobooks disappear soon after their young fledge, leaving all parental duties to males. Where did the females go? We radio-tagged a female and started to follow her as she fed her fledged young, but on the third night the radio signal disappeared. Maybe she had flown beyond the range of our receiver. We wandered through forest checking tree roosts one at a time and eventually found her roosting in a Native Cherry. Through binoculars we could see why the radio signal had failed. She had snapped off the antenna. This reduced the signal travelling distance to about 100 m, so we could not find her unless we were close. Sirtrack, the New Zealand makers of the transmitters had designed these radios for New Zealand Boobooks, or Moreporks, a smaller, weaker owl than Boobooks on mainland Australia. We contacted Sirtrack and they began reinforcing each antenna where it emerged from the transmitter, to make the transmitter robust enough to withstand the beaks of the more powerful Australian Boobooks.

We were unable to re-capture this female and change the broken radio. A second radio-tagged female left its breeding territory on Black Mountain soon after the young fledged and flew to a suburb near Parliament House, about 7 km away. There Steve Taylor began tracking her.

Over a 54-day period she was tracked for 12 nights, and ranged over about 122 ha for a 100 per cent MCP and 91 ha for 95 per cent MCP (see Figure 8.2). In this example you can see how two clusters of point locations indicate intense use (more foraging), one on the east side and one on the west side of the 95 per cent MCP.

Figure 8.1 Female Southern Boobook with bat. The harness on the attached radio-transmitter has broken at the designed weak point and soon will fall off.

It is important to sight the owls as often as possible instead of relying solely on transmitter signals. One night Steve went to the Cypress Pine she normally used as a day-roost. She wasn't there. Walking with the antenna over his head, Steve gained elevation, travelled to a hilltop and collected signals from a wider area, a standard technique when a signal is lost.

Steve located the signal in a large, two-storey home. Maybe she had roosted under the eaves or found her way into the attic. However, two hours later she had still not emerged.

The following night Steve stood outside and waited in the dark. Again she did not emerge.

The next day Steve knocked on the door and spoke to the elderly residents. The wife did the talking, her husband stood quietly by. They let Steve climb into the roof space and he found some pigeon feathers, and could still hear the signal but could not see the owl. The signal seemed to emanate from the chimney, and he wondered if the owl had become lodged there.

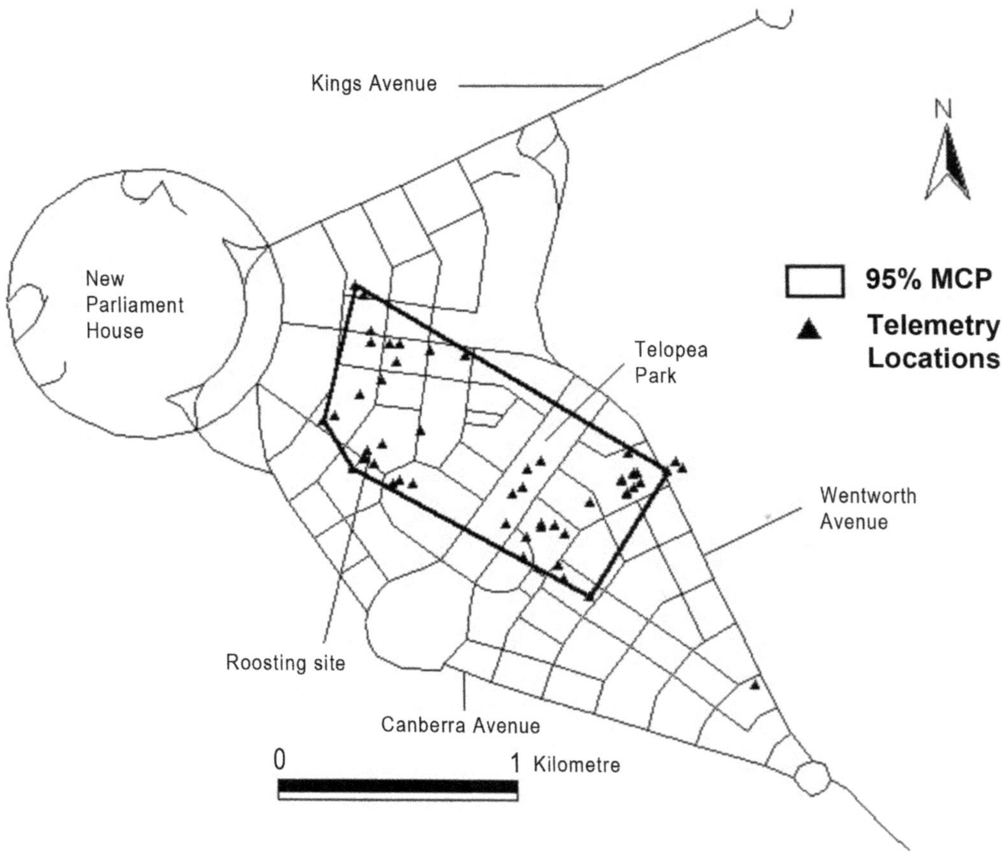

Figure 8.2 Winter home range of a female Southern Boobook in Canberra.

Steve visited the house again and spoke to the woman alone. Her husband lay napping in an upstairs bedroom. She mentioned that her husband had some form of dementia, and the hospital had attached a transmitter in case he wandered off.

The penny dropped. The owl never left the house because the elderly man had a transmitter frequency similar to the owl's, and it was he who never left the house.

Steve apologised to the woman and located the owl 1.5 km away. He followed her through winter. On 12 May 2001, seven years and four months after she was banded, Steve found her dead on her winter home range. A domestic cat had killed her as she hunted a backyard in the city.

Diet and hunting – medium-sized owls

Owls are designed to catch, subdue and dismantle live prey. They have powerful, clawed feet for catching and holding animals and hooked beaks for biting and tearing. Worldwide, there are species small enough to specialise on insects and large enough to capture mammals heavier than themselves. Some specialise in freshwater fish, caught in the manner of a fish-eagle. Owls catch prey by seizing it in their feet; they then kill it with a bite to the neck at the base of the skull. They have high metabolic rates and so have a high demand for food, and they need considerable skill to capture prey that is equally skilled at evading capture.

Medium-sized owls, such as the Southern Boobook and Long-eared Owl, hunt prey in somewhat different ways, and by exploring the diet and hunting methods of certain North American and European owls we can better understand diet and hunting in Australian owls.

9
Spotted Owls

The genus *Strix*, or 'wood owls', contains about 20 species spread across the world. These include the Spotted Owl, and the Great Grey Owl, the largest *Strix*. We have no *Strix* owls in Australia; the closest are on Java – the Spotted Wood Owl (not to be confused with the Spotted Owl in western North America) and the Brown Wood Owl. Owls in the genus *Strix* stop at Wallace's Line in Indonesia; we will say more about Wallace's Line in Chapter 41.

Spotted Owls are a well-studied species. The forests in western Washington State, Oregon, and California are the strongholds of the Spotted Owl *Strix occidentalis*. *Strix* in Latin (*strix*) and Greek (*strigx*) denotes a kind of owl, especially a strident one (according to Paul Johnsgard (2002) *striges*, the plural of *strix*, was applied in Rome to witches as well as to owls, as it was believed that witches could transform themselves into owls and suck the blood of sleeping children); *occidentalis* is Latin for 'western', as they occur in western North America. These confiding and gentle owls often roost next to a tree trunk in shadow and watch the researcher with eyes shiny black in the gloom. They are easily approached, even curious. Holding these owls, you are drawn to their dark, shiny eyes set in tawny half-circle discs in a brown face, and to the yellow-green beak between them.

Biologists recognise three races (subspecies) of Spotted Owl. Along the west coast of America, the largest race, the Northern Spotted Owl, lives from southern British Columbia down to northern California. From northern California down to the Mexican border lives the smaller California Spotted Owl. The smallest race, the Mexican Spotted Owl, lives east of the other two in the Rocky Mountains and deeply incised moist and cool canyons of southern Utah and Colorado south to the Guadalupe Mountains of west Texas and the Sierra Madre Mountains and Cerro de Tancitaro in Mexico.

Usually Northern Spotted Owls take mammals native to these different forests. They concentrate primarily on three mammals.

The Northern Flying Squirrel looks remarkably like a Sugar Glider, a favourite prey of Powerful Owls. It is interesting that two unrelated owls on opposite sides of the world, the Spotted and Powerful Owl, look so similar and prey on unrelated arboreal mammals that also look similar, an example of convergent evolution (where unrelated animals evolve to look and behave similarly). Flying Squirrels and Sugar Gliders are roughly the same size, but the Sugar Glider is a marsupial that carries pouched young, the Flying Squirrel a placental mammal that leaves small young behind in a nest when it goes out to forage. Like Sugar Gliders, Flying Squirrels are strictly nocturnal and, though common, they are not often seen by the public. They are the only nocturnal squirrels in North America. As with Australian gliders, they spread their feet to stretch their flight skin and glide long distances. Just before landing they drop their tail, lift their front paws and slacken the flight skin as an air brake, then pitch upwards at the end to land lightly with all four feet on the trunk of a tree. Flying squirrels have a habit of scurrying around to the other side of the tree just after they land, in case a predator, especially an owl, has spotted them. They eat seeds, nuts and birds' eggs, and store food in cavities and crevices for the winter. Their nest is built of shredded bark in a natural tree hollow or woodpecker hole, or they build an outside nest of leaves, twigs and bark.

Spotted Owls move through forest, sometimes along forest edges, stalking these little squirrels and other mammals, particularly the Dusky-footed Woodrat, the second important prey species, in southern Oregon and northern California. These woodrats live on the west coast from southern Washington to northern Mexico. They are greyish above, their bellies often washed with tan. They build 'lodges' on the ground or as high as 4.5 m up trees; the lodges can be 2.5 m across, and the nest for their young is buried deep inside. After mating, the males of this semi-colonial species live in separate tree nests. The Bushy-tailed Woodrat, another important prey species weighing some 250 g, has a wider distribution than the Dusky-footed Woodrat. It is found over much of western USA and British Columbia, and has a bushy tail like a squirrel. It is the original 'packrat'. These rats will even drop food if they spot a shiny spoon or coin, and pack it off to their house, sometimes built as high as 15 m up a tree.

There are no woodrats in Australia, but stick-nest rats live in drier regions and may be preyed on by *Tyto* owls. Placental rodents important to forest owls include the Bush Rat *Rattus fuscipes* and Broad-toothed Rat *Mastacomys fuscus* taken by Sooty and Masked Owls but seldom by Powerful Owls which specialise on arboreal mammals, not ground mammals.

Snowshoe Hares, a third species important for Spotted Owls, are also important for Lynx, Grey Wolf, Great Horned Owl, and Northern Goshawk. It is interesting that a medium-sized owl such as the Spotted Owl can lift heavy prey such as the Snowshoe Hare. They may carve Snowshoe Hares into pieces and deliver pieces to the nest, or probably most are juvenile hares. But, as mentioned above, owls also have a much higher lifting power than similar-sized falcons or hawks because of their broad, rounded, concave wings, and their ability to lift their wings so high over their heads. This could be one of the reasons why Powerful Owls in Australia can also lift such heavy mammals as Greater Gliders. However, Powerful Owls tend to ignore ground prey such as the European Rabbit, but *Tyto* owls and Barking Owls take them.

Biologists know that Spotted Owls concentrate on the above mammals because they find pieces of animal prey (*orts*) in Spotted Owl nests or under their day roosts. Because owls sometimes consume the heads of their prey away from the nest, or gulp small animals whole, biologists analyse regurgitated pellets (also called *castings*). These may be in the nest, under the nest or under the day roost some distance from the nest, for example where a male spends the day while his female incubates eggs or broods young. Biologists disagree about whether orts or castings best represent what an owl is eating. Sometimes a prey species shows up in castings that does not show up in orts, or a prey species shows up in orts that does not show up in castings, so many researchers combine orts and castings and take care not to double count. For example, if the teeth of 10 Flying Squirrels were counted in the castings, and the tails of eight flying squirrels were found under the nest, the biologist would count this as a Minimum Number of Individuals (MNI) of 10 Flying Squirrels, not 18.

A third set of evidence comes from observations, watching a nest through binoculars, or setting a camera in place to record prey deliveries. These three types of data – orts, castings, and observations – give a better estimate of owl diet than does any one data set on its own.

With orts and castings, biologists use two separate methods to estimate which prey species are important to an owl. In the first method, the biologist counts the number of bones or feathers, especially skulls, primaries (the large flight feathers,) feet and certain long bones, pieces of fur or mammal tails, that give an estimate of the *number* of a particular prey species found. For example, they might find orts and castings in a Spotted Owl nest that gave evidence for 17 individuals – 10 skulls of the Northern Flying Squirrel, 5 skulls of the Bushy-tailed Woodrat, and 2 skulls of the Snowshoe Hare. So counting this contribution of prey by number would give us:

Northern Flying Squirrel × 10 = (58.8%)

Bushy-tailed Woodrat × 5 = (29.4%)

Snowshoe Hare × 2 = (11.8%)

One would need larger samples to draw firm conclusions about Spotted Owl diet at this nest, but from this small collection a biologist might propose that Northern Flying Squirrels, at 59 per cent, are the most important prey by number, Bushy-tailed Woodrats at 30 per cent are second in importance, and Snowshoe Hares third.

A second way of estimating the importance of prey in a raptor diet is *biomass*, the actual amount of food represented by a prey item, the weight of meat if you like. How important is the prey species by weight, the actual amount consumed by an owl? Biologists calculate biomass because one prey species weighs more than another, and consequently, over time a large prey species could represent a larger percentage of the owl's diet than is represented by its number in the diet. Because a Northern Flying Squirrel weighs about 170 g, a Bushy-tailed Woodrat weighs 400 g, and a Snowshoe Hare 1500 g, a biologist could calculate the importance of each of these three prey species in the example above by multiplying the number of each by the species' weight:

Northern Flying Squirrel: 170 g × 10 = 1700 g (25%)

Bushy-tailed Woodrat: 400 g × 5 = 2000 g (30%)

Snowshoe Hare: 1500 g × 2 = 3000 g (45%)

This gives a different estimate of the importance of Northern Flying Squirrel in the diet compared with Snowshoe Hare. It also gives a total of 6700 g from 17 prey items, and a mean (average) prey weight prey of 394 g. Biologists use another factor called wastage to measure the percentage of a large animal that predators waste, that is, throw away because they cannot eat it all in one sitting. We do not include wastage factors in our discussion here.

So, in the example above, they found evidence of only two Snowshoe Hares in this nest, but they were the most important prey by weight, or biomass. As is often the case, a larger sample can change these findings. In studies with larger sample sizes, the most important prey species in terms of biomass for the coastal Spotted Owl is the Northern Flying Squirrel.

Spotted Owls concentrating on woodrats have smaller home ranges than Spotted Owls concentrating on flying squirrels, perhaps because the ground-dwelling woodrats live at higher densities than the tree-dwelling flying squirrels, or maybe woodrats are more vulnerable to attack than are flying squirrels. In some cases, as the amount of older forest decreases good Spotted Owl habitat, the home range size appears to increase. Besides flying squirrels and woodrats, Spotted Owls also capture mice, voles, brush-rabbits, insects, amphibians and birds including small species of owls.

10

Boobook summer diet

Southern Boobooks hunt insects and birds, mainly in the forest canopy, grass or in bushes. At sundown one male would leave his day roost, some 250 m from his incubating female, and hunt low bushes on his way to the nest, then meet her for their first contact that night. He crashed into low foliage after small passerines like pardalotes and fairy-wrens that had gone into cover at dusk, making a crack when he pitched into the bushes. He often caught a bird, continued toward the nest and delivered the body to his incubating female.

Boobooks also snatched spiders off tree limbs, perched on a limb and stripped off bark to peer underneath or caught insects such as moths in mid-air. They sat in a treetop and watched the night sky, then flew out to intercept moths, flipping backwards and shooting out their feet to snare one. They caught bats in the same way, pulled stick insects from the outer foliage of eucalypts, often hanging upside down to do so. They took centipedes and grasshoppers from the ground and tended to hunt more invertebrates on warm nights when invertebrates were moving and visible. On cold nights, they caught more birds and mammals. Riegert and Fuchs (2004) found the same thing with Common Kestrels in Europe. Mean temperature was positively correlated with the abundance of insect prey found in their diet.

Observations can give some idea of what owls eat, but a more thorough analysis comes from adding prey remains and owl castings that owls shake out of their throats at dusk and drop onto leaf litter. In winter these pellets looked like smaller versions of the Spotted Owl pellets mentioned above, oblong packages of grey felt, packed with fur, bone and feathers and the size of a joint on your thumb (see Figure 10.1).

Figure 10.1 Southern Boobook casting.

In summer things can be difficult. Australia is ant-rich and most owl researchers struggle with ants in their summer prey studies. In his book, *The Future Eaters*, Tim Flannery said that more species of ant live on Black Mountain in Canberra, the site of our study, than in all of Great Britain. Summer Boobook castings differ from a lot of owl castings because the Boobooks eat many insects; the castings reflect shiny transparent wings and beetle casings. Rivers of ants mill around under the day roosts, and the ants break up the castings and carry away pieces before they can be found and taken away for analysis (see Figure 10.2). Rohan Bilney experienced this problem in his study of Sooty Owls in Victoria. The solution is to collect the pellet as soon as possible after the owl casts, and before insects dismantle it, though some researchers use ant killer in a small circle of ground underneath the perch.

Tony Rose, an expert at analysing pellets, picks apart each casting to determine what the Boobooks, Barking Owls, Barn Owls or Powerful Owls have eaten. Tony, a retired dairy farmer living in Forster, New South Wales, has conducted many studies and published widely. He is self-taught, has a prodigious knowledge of natural history and the capacity for hours of detailed work on Boobook prey. Tony has a private collection of insects, spiders, birds, mammals and reptiles to crosscheck against the pieces of animal we found in the Boobook castings. He could identify mammal hairs, bird beaks, legs, feathers, and spider mouthparts of the animals that Boobooks prey on, and estimate Minimum Number of Prey

Figure 10.2 Ants at a just-dropped Southern Boobook casting.

Items (MNI) for each collection made in the field. This means the owls caught *at least* these species, in these numbers and probably more.

Tony would break the castings apart and compare the contents with those in his collection or with the collection held at the Australian Museum in Sydney (see Figure 10.3). The work was tedious, and much of it was done through a microscope.

As detailed in Chapter 9 with Spotted Owls, researchers can pool three sources of prey data: (1) animals that researchers saw them catch, (2) animals identified from pellets, and (3) animals identified from prey remains or orts dropped near the nest or roost. Tony took care to minimise biases. For example, he did not assume that each pellet containing pardalote feathers should be counted as one pardalote in the list of prey, because mated Boobooks shared prey items, with each other and their young. Three pellets under a perch all containing pardalote feathers could be three owls sharing one pardalote. Tony counted the number of legs and beaks, or primary feathers, to determine if the pardalote feathers in the three castings meant three different pardalotes, or just one.

In the initial study of three nests in 1993 and 1994 (Table 10.1), Tony found that most prey items by number were invertebrates (60.9 per cent). Among these, beetles were most common (39.1 per cent), followed by moths and butterflies at 10.9 per cent. As explained above, you get a different picture when you consider the weight of each prey species and calculate biomass. For example, a Crimson Rosella weighs

Figure 10.3 Tony Rose comparing prey found in Southern Boobook pellets to species in his invertebrate collection.

about 131 g and these Boobooks caught three. Their importance by number was 6.5 per cent, but their importance by biomass was 32.1 per cent. Birds provided 89.5 per cent of the dietary biomass for breeding Boobooks on Black Mountain and Aranda Bushland. The main birds they ate were Common Starlings (24.5 per cent) and juvenile Crimson Rosellas (32.1 per cent). This cast doubt on the conventional wisdom that Southern Boobooks were small, insectivorous owls.

In this small sample Tony added four new species to the known diet for Southern Boobooks – Crimson Rosellas, Striated Pardalotes, a robin that Tony could not identify to species and Black-faced Cuckoo-shrikes. Though some writers, like König *et al.* (1999), suggested that Southern Boobooks caught birds up to the size of House Sparrows, these owls took Crimson Rosellas, and we thought the larger females caught them. But years later, watching a pair on Black Mountain in 2004, a male left his fledged brood at dusk just as the young were starting to stretch and trill and preen each other, getting ready for the night. The male stalked a roosting Crimson Rosella, flying silently from tree to tree. He captured the rosella as it squawked and fought while the Boobook hung upside down from eucalypt leaves, as they do when they catch invertebrates. He tried to subdue it,

Table 10.1 Number and percentage of prey species, prey items and biomass from three Southern Boobook nests in 1993 and 1994

Prey species		Weight (g)	n	% by number	% by mass
Mammals					
House Mouse[i]	*Mus musculus*	18	2	4.3	2.9
Bat sp.	Microchiroptera	20	1	2.2	1.6
Birds					
Crimson Rosella – juvenile[n]	*Platycercus elegans*	131	3	6.5	32.1
Spotted Pardalote[n]	*Pardalotus punctatus*	9	1	2.2	0.7
Striated Pardalote	*Pardalotus striatus*	12	1	2.2	1.0
Robin sp.[r]	Petroicidae	15	1	2.2	1.2
Black-faced Cuckoo-shrike[n]	*Coracina novaehollandiae*	105	2	4.3	17.1
Common Blackbird[i]	*Turdus merula*	95	1	2.2	7.7
Common Starling[i]	*Sturnus vulgaris*	75	4	8.7	24.5
Unidentified		63	1	2.2	5.1
Reptiles					
Skink sp.	Scincidae	15	1	2.2	1.2
Invertebrates					
Wolf spider	Lycosidae	3	1	2.2	0.2
Spider	Araneae	3	1	2.2	0.2
Christmas beetle	*Anophlognathus* sp.	2	6	13.0	1.0
Scarab beetle	Scarabaeidae	2	1	2.2	0.2
Beetle	Coleoptera	2	11	23.9	1.8
Bug	Hemiptera	2	1	2.2	0.2
Moth/butterfly	Lepidoptera	2	5	10.9	0.8
Praying mantis	Mantodea	2	2	4.3	0.3
Total			**46**	**100**	**100**

[i]Introduced species.
[n]Species not previously reported (Higgins 1999; Penck and Queale 2002).

flapping his wings and dislodging the rosella from the tree. The male crushed the base of the rosella's skull with his beak and then presented the carcass to his mate.

Bird killing was limited to adults, both female and male caught both vertebrates and invertebrates, but the fledged young never caught vertebrates. Juveniles caught beetles and other insects. The same trend was described for Eurasian Kestrels by Riegert and Fuchs (2004). Breeding adult kestrels brought in vertebrates, but their young caught only insects. Adult Boobooks brought birds, reptiles and mammals to the nest and to their fledged young, so the young knew these prey, and probably caught them in winter, but caught insects and spiders on hot summer nights (see Figures 10.4 and 10.5).

Figure 10.4 Juvenile Southern Boobooks took only invertebrates, while adults took more birds.

This first analysis of Boobook breeding diet near Canberra was small. Larger samples of summer prey were needed, and samples of winter prey for Tony to analyse and compare. Maybe they would tell a different story.

Figure 10.5 Southern Boobooks take many spiders, such as this Canberra Huntsman.

11

Boobook prey size

Sugar Gliders and possums

Sugar Gliders are commonly found in Canberra woodland with Southern Boobooks. They weigh about 120 g, similar to a parrot's weight; however, the Boobooks seldom took Sugar Gliders. They treated them as enemies, not food. Something about the aggressive behaviour of glider family groups puts the Boobooks on the defensive, and makes the gliders immune to frequent attack as a prey species. Sugar Gliders do not seem to fear the owls as much as they should, and David Lindenmayer from the Australian National University said that Sugar Gliders give a certain call that brings in the rest of the group to mob a Boobook.

Even if they are close to an owl family, Sugar Gliders will climb out onto the end of a thin twig and yap, or sail from a treetop across a gap in the canopy. Sometimes the owls fly after them, clacking their beaks in warning. This clacking is like two pieces of wood clapped together. Some observers believe Boobooks make this clacking sound with their tongue, but others say it is mandibles clacking together. There are reports of Boobooks attacking Ringtail Possums as prey (McNabb 2002), but these Boobooks were beak clacking as warning, not gliding in silently for a kill. This signals an attack on an enemy, not hunting. The owls often follow the glider through the air and try to knock it to the ground. When the owls strike a glider, they strike with talons wide open, as when capturing prey. When the owl attacks a glider clinging to a tree, the glider often drops to the ground and escapes. But sometimes a Sugar Glider will dodge around the trunk of a tree, persist and refuse to leave a patch of trees owned by the owls (see Figure 11.1). Yellow-bellied Gliders behave in a similar fashion with Powerful Owls.

Figure 11.1 Sugar Gliders are common in woodland; however, Boobooks rarely take them.

It is easier to understand why Boobooks commonly attack Ringtail and Brushtail Possums when these mammals crawl down trunks or onto the ends of branches close to an owl's nest. Both the male and female Boobooks attack possums, making the beak-clacking noise. Boobooks sometimes dislodge possums and they fall with a thud to the forest floor. Brushtail Possums weigh about 10 times as much as a Boobook (see Figure 11.2), but female Boobooks in particular can strike them in the hindquarters and make them lose their grip, especially possums clinging to a smooth-barked trunk and facing downwards. Brushtail Possums are omnivorous and eat eggs. They are big and aggressive enough to confiscate a nest hollow used by Boobooks (Figure 11.3). Perhaps Boobooks don't discriminate between omnivorous Brushtails and fruit-eating Ringtails, or perhaps suitable hollows are in such short supply that all hollow-nesting birds and animals are deterred.

Comparing prey size with the size of the raptors that catch them

There is, then, a certain size range for Boobook prey. Possums are too large, but smaller animals like mice and rats or small birds are optimal. With prey data from raptors, a biologist can estimate how big the average prey is compared with the body weight of the owl, and how this compares with other raptors. Because one very large, or one very small prey item can distort a mean (average) prey weight, biologists use Geometric Means to correct this, a log-transformation of the mean

Figure 11.2 Southern Boobook attacking Common Brushtail Possum that was attempting to raid its nest.

Figure 11.3 Brushtail Possums enter Southern Boobook nests if they can to usurp the hollow or consume eggs and young.

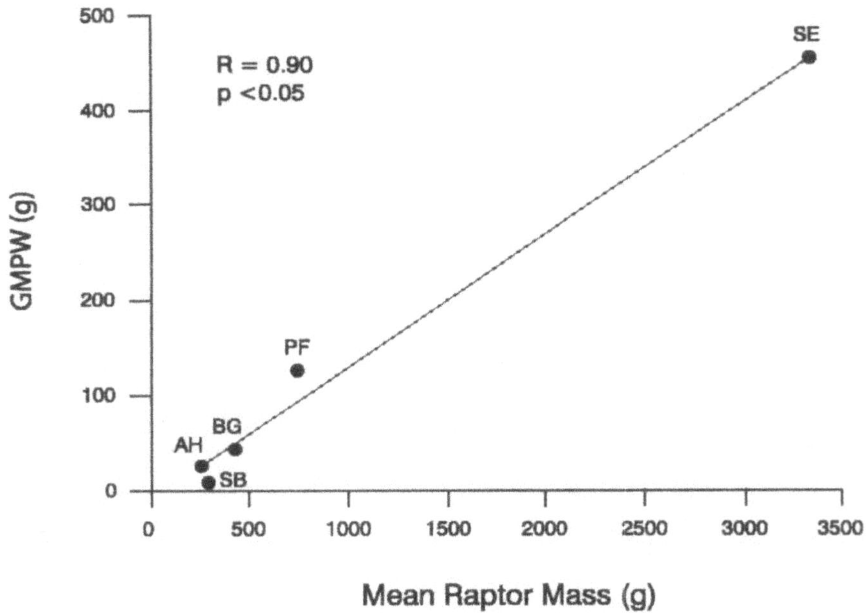

Figure 11.4 Correlation between predator and prey size, given by the Mean Raptor Weight and the GMPW respectively: White-bellied Sea-Eagle (SE), Brown Goshawk (BG), Australian Hobby (AH), Peregrine Falcon (PF) and Southern Boobook (SB) – dotted trendline added for perspective only. From Olsen *et al.* (2006).

weights of individual prey species, before calculating the grand mean weight. Using these Geometric Mean Prey Weights (GMPW) you can calculate a ratio of the weight of the prey to the weight of the raptor, and suggest which raptor takes large prey in relation to its size and which takes smaller prey. Because raptors can be dimorphic, most females are heavier than males, researchers use an average weight of both sexes for each raptor species (see Figure 11.4). Among five Australian raptors there was a strong positive correlation between raptor weight and prey weight. Heavier raptors took heavier prey, though Southern Boobooks took more small prey relative to their size, about 3 per cent of their own body mass. Peregrines, on the other hand, took more large prey compared to their size, about 18 per cent of the Peregrine's body weight.

Two owls from the Northern Hemisphere were added to the analysis, both about the weight of Southern Boobooks. These were mammal-specialists from colder climates – the Long-eared Owl and Northern Hawk-Owl (Table 11.1). Southern Boobooks took proportionally smaller prey than the similar-sized Long-eared Owl and Northern Hawk-Owl, probably because temperate-climate Boobooks take more insects than do similar-sized, cold-climate Northern Hemisphere owls.

Overall, owls take prey that is proportional to their body weight; however, there are exceptions and we will discuss some of these in Chapter 18.

Table 11.1 GMPW, predator weight and prey/predator ratio for five species of raptor breeding in the Canberra area between 1990 and 1994 (ordered by decreasing GMPW), and for two Northern Hemisphere owl species

Predator	GMPW (g)	Predator weight (g)	Prey/predator ratio
White-bellied Sea-Eagle	470.7	3300	0.14
Peregrine Falcon	132.0	745	0.18
Brown Goshawk	44.9	440	0.10
Australian Hobby	20.8	250	0.08
Southern Boobook	6.9	283	0.02
Long-eared Owl	21.1	255	0.08
Northern Hawk-Owl	24.9	295	0.08

12

Northern Hawk-Owls and Long-eared Owls

In Table 11.1 the two Northern Hemisphere owl species, the Northern Hawk-Owl and Long-eared Owl, are similar in weight to Southern Boobooks, but take larger prey. These two owl species and Southern Boobooks have different body shapes and hunting strategies, different average prey size and clutch sizes. Two of them are nocturnal, one hunts in daylight. One often nests in woodpecker holes, one uses naturally formed tree hollows in eucalypts and one prefers old crows' or magpies' nests in trees. Some of these differences and similarities are explored below.

Northern Hawk-Owls

In clean early sunlight on Highway 1 east of Winnipeg, Canada, the scenery changed from blank snowfields to naked scrubby deciduous trees and small conifers about 10 m high. It was –20°C. In summer, this Shield Country is dotted with lakes, black flies and mosquitoes. According to Jim Duncan, the Canadian owl expert, this makes work on Northern Hawk-Owls and other species uncomfortable, even torturous. In winter the lakes and mosquitoes are sealed below ice and snow and the owls are easier to watch.

A Northern Hawk-Owl *Surnia ulula* was perched on top of a snowy conifer, sitting more horizontally than most owls do and not owl-shaped, but not quite Peregrine-shaped either. It had a pot-bellied look, like a sitting Brown Falcon.

Biologists are uncertain as to the origin of the genus name *Surnia*, but it may come from the modern Greek *surnion*, a vernacular for the Tawny Owl. The Northern Hawk-Owl is the only species in its genus and the species name *ulula* comes from the Latin – 'crying out in pain'. This may relate to the Greek *alala* and *ololuge*, 'to howl, especially in pain'. They breed in boreal zones of Canada and Eurasia, coniferous or mixed coniferous-deciduous forest, burned-over areas and muskeg in a belt circling the world. They are one of the least-studied birds in North America.

Northern Hawk-Owls weigh a little more than Southern Boobooks – about 300 g for males, 345 g for females, and they have longer tails, 176 mm compared to 135 mm for Boobooks, so they look bigger than a Boobook. They have striking black and grey plumage. Facing you, a black mark curves down each side of their face, as if the eyes are parenthesised with a black marking pen. Some have another thick black mark across their throats.

Unlike Southern Boobooks, and most other owls, Northern Hawk-Owls are diurnal. They hunt in bright sunlight. They have other non-owl characteristics, resembling accipiters (goshawks) and falcons, a more flat-topped head and a smaller facial disc. While most owls have rounded wings, Hawk-Owls have wings that taper to a point. Most owls, such as the Southern Boobook, have sound-deadening filaments at the edges of their wing feathers, for silent flight. These are poorly developed in Northern Hawk-Owls. The long, graduated tail tapers to the tip and the bright yellow eyes are comparatively small, almost beady.

The Hawk-Owl on Highway 1 scanned the open woods for 15 minutes then arrowed down from the top of the tree, swift and direct, wings flickering like the wings of a falcon, chasing a small bird over the snow, using its long tail as a rudder. In the breeding season Hawk-Owls catch voles (Microtidae), lemmings and other mammals, about 76 to 96 per cent of the biomass taken, but they take many Snowshoe Hares in Canada (Rohner *et al.* 1995).

The average clutch size is large compared with Boobooks and other Australian raptors. In one study it was 6.3; the Southern Boobook lays about three eggs. Northern Hawk-Owl eggs and juveniles must suffer high mortality, and adults may not survive for as many years as adult Boobooks do.

Unlike North America and Eurasia, Australia has no 'primary excavators', such as woodpeckers that drill holes in trees. Woodpeckers are crucial to the biology of many North American owl species, but Australian *Ninox* are limited to naturally formed tree cavities. Northern Hawk-Owls favour a nest hole made by species such as the Pileated Woodpecker, or they use the top of a broken-off stump or an old bird's nest. In Europe, where few large trees remain, they use nest boxes.

In winter Northern Hawk-Owls may expand their diet from voles and other mammals to more birds, even Willow Ptarmigan (the Red Grouse in Scotland) and Hazel Grouse. Willow Ptarmigan weigh about 550 g and Hazel Grouse some 400 g, both larger than the average Northern Hawk-Owl. These owls can irrupt south from their normal range into southern latitudes when populations of their mammal prey crash. Australian *Ninox* do not irrupt in this manner, but Australian mammal specialists in the genus *Tyto*, such as the Barn Owl and Grass Owl, do.

Long-eared Owls

Pullman, Washington, lies in the Palouse, a region of steep, rolling hills and grassland, green crops and Hawthorn bushes in summer, tawny-coloured fields in autumn. In winter, black fallow soil is streaked with snow. The Nez Perce Indians bred Appaloosa horses and fought the US Army there, and this was where they led the most famous retreat in Native American history. These grasslands may be the richest agricultural lands in the United States, almost all of it is cultivated and there are almost no native plants. However, Long-eared Owls *Asio otus* are common in the Palouse because the land is rich in small mammals.

The genus name *Asio* comes from the Latin for 'a horned owl' because of the feather tufts on their heads. According to König and Weick (2008), the genus *Asio* has eight species spread across North and South America, Europe, Asia, Africa, Madagascar and quite a few islands. Pliny reportedly named the Long-eared Owl for its 'ass-like' appearance, due to its long 'ears'. The species name *otus* comes from the Greek *otos* or 'eared' (Johnsgard 2002). These ear tufts or 'horns' are usually invisible in flight.

Long-eared Owls are larger in dimensions than Southern Boobooks, but this is mostly due to their feathers. Wing-chord length (from the carpal joint to the end of the longest primary feather) is about 285 mm for Long-eared Owls and about 240 mm for Boobooks. The tail is about 146 mm for Long-eared Owls and 135 mm for Boobooks. Male Long-eared Owls weigh about 245 g, females about 279 g (Earhart and Johnson 1970), close to the weight of Boobooks. They have reddish-tinted plumage, with darker streaks on their front, a mottled brown back and a large head with conspicuous ear tufts that no Australian owls have. They have a buff-coloured facial disc with yellow eyes, and contrasting light facial bristles around the beak and white 'eyebrows' that many nocturnal owls have, including Southern Boobooks, probably used for signalling.

Long-eared Owls are said to range through North America including Mexico, and through Eurasia as well as the Canary and Galapagos Islands. All these owls look like *Asio otus*, but it is likely that some will be given species status, especially those on islands. Some argue that the North American and Eurasian forms are two different species.

The range of Long-eared Owls overlaps the range of Northern Hawk-Owls, but Hawk-Owls are more northerly, Arctic and subarctic breeders. Unlike Northern Hawk-Owls, Long-eared Owls have silent flight and hunt mostly at night, like

Boobooks. Unlike Northern Hawk-Owls and Southern Boobooks, Long-eared Owls have large, asymmetrically placed ear openings on the sides of their heads that help them triangulate the sound of moving prey. They locate much of their prey by ear, and can capture mice in complete darkness (Marti 1974).

They hunt exclusively in open habitats and their major prey according to Marti (1976) and Marks (1984) is a wide variety of small mammals weighing 10–100 g. Mean prey mass varied from 25 to 45 g, especially Meadow Voles, Prairie Voles and Montane Voles (mammals weigh more than birds and insects hence the larger predator/prey ratio in Table 11.1). Their long rounded wings and low mass gives them low wing loading and considerable buoyancy. Like a harrier, they quarter low over grassland or stubble, about 50 to 150 cm off the ground and seem to bounce with each flap, then suddenly stall, drop with their talons spread and pin a small animal to the ground. Their short toes are made for capturing mammals; they do not capture many insects in mid-air as Southern Boobooks do with their longer toes.

Long-eared Owls live close to short vegetation for suitable hunting, but close to wooded vegetation for roosting and nesting. Unlike Northern Hawk-Owls, they spend the daylight hours perched near tree trunks in dense vegetation, like conifer woodland, where they are cryptic and nearly invisible. When disturbed they slim their body down like a Tawny Frogmouth and erect their ear tufts to look more like part of the tree.

Clutch size in North America is about 4.5 eggs, larger than Australian *Ninox* and smaller than Northern Hawk-Owls. Unlike Boobooks and Northern Hawk-Owls they usually do not nest in tree hollows. They prefer stick nests, especially corvid nests in pines, but they also use the roofed nests of Black-billed Magpies *Pica pica*, in hawthorn bushes *Crataegus.* They also use hawks' nests. This may make their young more prone to predation. One Palouse nest made by Cooper's Hawks in a second-growth fir tree had two owl nestlings. They stood fluffing their feathers and snapping their beaks at us. The later arriving Cooper's Hawks killed and ate these nestlings. Long-eared Owls are the most preyed upon owl in Europe and they are commonly found as prey in the nests of Northern Goshawks, Eagle Owls, Common Buzzards, Peregrine Falcons and other raptors. This is probably because, in the far northern parts of their range, this predominantly nocturnal owl is forced to hunt in twilight or during the day, leaving them more exposed to diurnal predation. In contrast, Southern Boobooks and other Australian owls are seldom taken as prey.

Many Long-eared Owls migrate long distances in the non-breeding season, and, unlike most owls, they often roost communally. Australian *Ninox* do not do this, but Australian Barn Owls and Grass Owls do specialise in terrestrial mammalian prey and sometimes do roost communally. Each day Long-eared Owls leave the communal roost at dusk and hunt independently through the night. They appear to

Figure 12.1 Just-fledged Long-eared Owl in hawthorn bush.

take similar prey in winter as in summer – small mammals, as do Eastern Barn and Grass Owls. Eastern Barn Owls prey mainly on small terrestrial mammals, primarily small rodents, 61–99 per cent by number in a summary by Higgins (1999), and in inland Australia they also take dasyurid marsupials. For example, Debus *et al.* (2004) found that Barn Owls near Lake Frome in South Australia took 67 per cent mammals, including 4 per cent Fat-tailed Dunnart *Sminthopsis crassicaudata* and 62 per cent House Mouse. Eastern Grass Owls prey mainly on small terrestrial mammals, especially Muridae such as Long-haired Rat *Rattus villosisimus*, Canefield Rat *Rattus sordidus* and Pale Field Rat *Rattus tunneyi*, but also small dasyurid marsupials, occasionally small gliders and some birds and insects (Higgins 1999). As with Long-eared Owls, Grass Owls hunt by quartering over the ground, much like a harrier, and making quick plunges into the grass. Australian *Ninox* and the two Australian forest *Tyto* do not regularly quarter.

In Chapter 13 we contrast the diet and hunting of medium-sized North American owls with the winter diet in Southern Boobooks.

13

Winter Boobook diet

Several texts and at least one research paper claim that Southern Boobooks take mainly vertebrates, not insects. Penck and Queale (2002) analysed crop and stomach contents from 117 Southern Boobooks in South Australia, mainly from road-kills and window-strikes in autumn and winter. They concluded that invertebrates made up 95.9 per cent of the diet. Estimates from pellets (castings) alone, they said, might underestimate the number of invertebrates with no hard, indigestible parts that would persist in pellets.

A problem with this claim was that researchers analysed crop contents. Owls differ from other raptors, like eagles or hawks, because they do not have crops. Penck and Queale should have, instead, been analysing gizzard contents, so one has to wonder if samples were confused with other road-killed birds that do have crops, or confused crops with gizzards.

The gastric juices in a hawk gizzard, where pellets are formed, can be about six times as concentrated as gastric juices in an owl gizzard. Consequently, an owl casts a pellet with whole animal bones and skulls and hard insect parts in it, whereas a hawk might cast a pellet looking more like a ball of felt because small bones were digested. Why didn't Peake and Queale find more vertebrate bones in the road-killed Boobooks they analysed? Ed McNabb (2002), as well as James Fitzsimons and Tony Rose (2007), both said that wintering Boobooks in Victoria preyed mainly on mammals. Who was right?

In an earlier study, Tony Rose (1996) analysed both stomach contents and pellets from Southern Boobooks. He found proportionally more invertebrates in stomachs, and proportionally more vertebrates in pellets. This seemed to support claims made in Penck and Queale. Penck and Queale used Tony's (1996) study, and others, to support their argument that estimating diet from pellets alone would underestimate the number of Lepidoptera (moths) and Araneae (spiders) in the diet – hard-bodied prey is regurgitated in pellets, but soft-bodied prey leave no indigestible material (see Figure 13.1). But Tony Rose had suggested something a little different. He said the discrepancy between stomach (gizzard) samples from dead owls, and pellet samples from live owls may signify:

1) pellets made up of spiders, beetles, moths and other invertebrates breaking down more quickly *in the environment* so they are not found by collectors, particularly when ants got to them first
2) pellets containing more vertebrate remains, bones, fur and feathers collected by people because they look more interesting, and
3) many road-killed owls are juveniles forced into marginal habitats where they subsist on invertebrates.

Compared to breeding adults, juveniles take more invertebrate prey and more of them are hit by cars.

Tony's argument showed a clear understanding of owl behaviour – those killed by cars and window-strikes might be inexperienced juveniles. These juveniles often

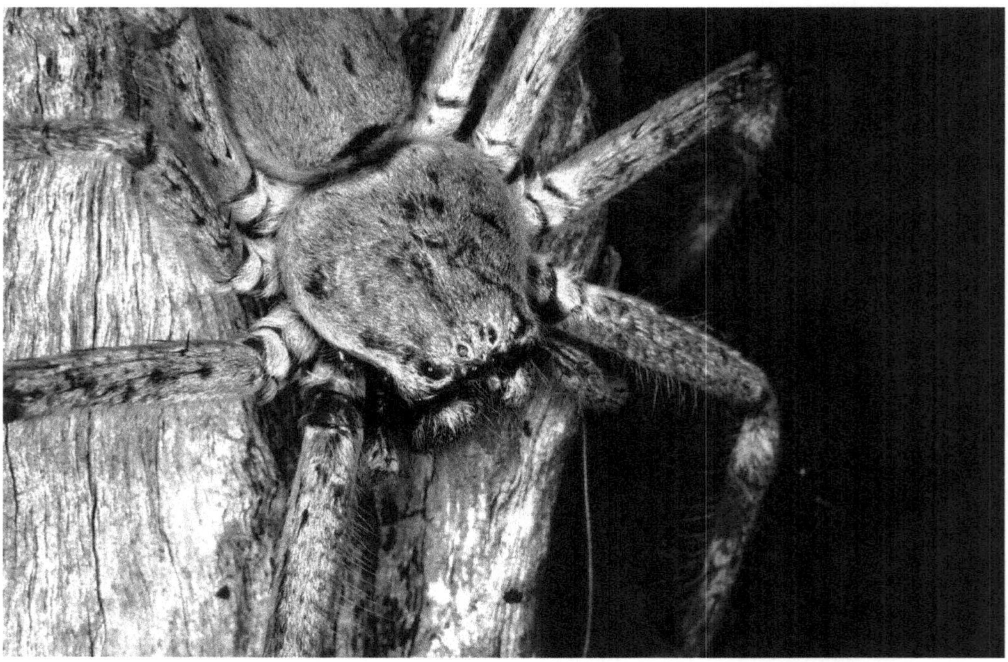

Figure 13.1 Some researchers believe that soft-bodied invertebrates, such as Huntsman spiders, are underestimated in prey analysed from castings.

Table 13.1 Number (n) and percentage of prey items and biomass from roosts in nine Southern Boobook territories 1997–2005

Prey species		Weight (g)	n	% by number	% by mass
Mammals					
House Mouse[i]	*Mus musculus*	18	83	16.7	28.4
House Mouse juvenile	*Mus musculus*	9	71	14.3	12.1
Black Rat	*Rattus rattus*	180	3	0.6	10.3
Black Rat juvenile	*Rattus rattus*	50	8	1.6	7.6
Bush Rat	*Rattus fuscipes*	125	1	0.2	2.4
Gould's Wattled Bat	*Chalinolobus gouldii*	14	1	0.2	0.3
Subtotal			**167**	**33.6%**	**61.1%**
Birds					
Red-rumped Parrot[n]	*Psephotus haematonotus*	61	1	0.2	1.2
White-browed Babbler	*Pomatostomus superciliosus*	39	1	0.2	0.7
Red-browed Finch	*Neochmia temporalis*	11	6	1.2	1.3
House Sparrow[i]	*Passer domesticus*	27	5	1.0	2.6
Common Myna[i]	*Acridotheres tristis*	116	3	0.6	6.6
Common Blackbird[i]	*Turdus merula*	95	1	0.2	1.8
Common Starling[i]	*Sturnus vulgaris*	75	12	2.5	17.1
Birds undetermined		63	4	0.8	4.8
Subtotal			**33**	**6.7%**	**36.1%**
Invertebrates					
Huntsman Spider	Heteropodidae	0.5	14	2.8	0.13
Wolf Spider	Lycosidae	0.5	73	14.8	0.7
Spider undetermined	Araneae	0.5	11	2.2	0.1
Grasshopper	Order Orthoptera	0.5	47	9.5	0.44
Locust	Order Orthoptera	0.5	7	1.4	0.07
Cricket	Order Orthoptera	0.5	8	1.6	0.08
Gryllids	Order Orthoptera	0.5	6	1.2	0.06
Eucalyptus Longhorned Borer	*Phoracantha semipunctata*	0.5	18	3.6	0.17
Longicorn Beetle	Cerambycidae	0.5	4	0.8	0.04
Scarab Beetles	Fam. Scarabaeidae	0.5	4	0.8	0.04
Weevil	Curculionoidae	0.5	1	0.2	0.01
Other Beetles undetermined	Order Coleoptera	0.5	41	8.3	0.4
Cockroach	Blattodea	0.5	9	1.8	0.08
Moth undetermined	Order Lepidoptera	0.5	44	8.9	0.4
Insects undetermined		0.5	9	1.8	0.08
Subtotal			**296**	**59.7%**	**2.8%**
Total			**496**	**100**	**100**

[i]Introduced species.

[n]Species not previously reported (Higgins 1999; Penck and Queale 2002; McNabb 2002).

Figure 13.2 Boobooks took prey such as Marbled Geckos but sometimes these did not show up in castings or prey remains.

wander. They hunt in new areas and are more at risk in strange settings from collisions with cars or window-strikes. And juveniles should take proportionally more invertebrates than breeding adults take.

In a sample collected from Boobooks wintering in Canberra from 1997 to 2005 (Table 13.1) Tony Rose identified that vertebrate prey dominated by biomass. Both the analyses of stomach contents from vehicle- and collision-killed owls and the analyses of pellets and prey remains from under the nests and roosts of breeding owls have biases. Pellets in the above study were collected at dusk, just after the owls dropped them and before ants and rain could break up perishable pellets made up of insects. For breeding data, photographs of prey deliveries showed that certain vertebrates brought to the nest, like geckos and rosellas, did not show up in pellets the following day (see Figure 13.2). It seems that gecko bones were completely dissolved, and a reptile expert, Joe McAuliffe from the National Botanic Gardens in Canberra, said that pythons also dissolve gecko bones so these do not show in the droppings. Maybe the Boobooks pick carefully through the meat of rosellas and swallow no bones or feathers, though sometimes rosella feathers did show up.

The diet of Canberra Boobooks in summer and winter was, by weight, mainly birds and mammals, not insects, and the combination of three collection methods – orts, pellets and direct observations – contributed to the picture. They took more invertebrates than did Northern Hawk-Owls and Long-eared Owls discussed above, so their diet is more varied and predator/prey ratio (Table 11.1) is lower.

14

Great Horned Owls and versatility

In North and South America the Great Horned Owl is ubiquitous (see Figure 14.1). They are the size of a Powerful Owl and as widespread as the Southern Boobook is in Australia. Why is the Great Horned Owl so widespread while the similar-sized Powerful Owl is limited to coastal and subcoastal Australia?

The day temperature in Winnipeg rose to –10°C, and locals used the break in cold weather to picnic. One family drove to the park. With a flip of their mittens, they dusted snow off the wooden seats and tables, spread out chicken, bread, a thermos of coffee, canned drinks for the kids and ate lunch in the wintry sunshine.

A minibus from the owl conference travelled on the plains outside Winnipeg to Oak Hammock Marsh, a wetland for migratory waterfowl now covered with ice. With binoculars on their laps, they rubbed clear spots in the foggy windows to see their first Winnipeg owl. Four owls are common there in mid-winter – the Great Horned Owl, Snowy Owl, Great Grey Owl, and Northern Hawk-Owl. White countryside was broken by occasional clumps of naked grey willow and dark conifers. A long-tailed magpie dipped low over the snow and skittered along the top of a wooden fence. In a mixed bunch of deciduous and conifer trees sat a Great Horned Owl. It sailed out of the tree and pitched up in a conifer, apparently hunting in broad daylight.

The Great Horned Owl's scientific name, *Bubo virginianus*, is derived from the Latin *Bubo*, for a horned or hooting owl, probably in reference to its low-pitched hooting call, but maybe from the Greek *buzo* 'to hoot' and *buas* 'horned owl'. The species name, *virginianus*, is Latin for Virginia, a southern state in the US.

Figure 14.1 The Great Horned Owl is a powerful predator, the size of a Powerful Owl. Researchers in Arizona use this tame owl to capture nesting Cooper's Hawks (photo by Addie Mannan).

Through binoculars the owl looked commanding, its iris bright chrome yellow, its body mottled with dusky and whitish spots on brown folded wings. Perched in bright sunshine it had dark horizontal pencil lines across its breast. These owls use their bright white throat and white eyebrows as signals of dominance when they call at night. Wing primaries were ground colour splashed with ochre. Sitting with its eyelids drooping, its cat-like 'ears' sticking up, it was frightened of no other bird on the Winnipeg plains. Great Horned Owls are common in summer and in winter. They live only in the Western Hemisphere, though they may form a superspecies with the Eurasian Eagle Owl. Great Horned Owls range from the tip of South America north through Central America, the

US, and Canada to the Arctic. Unlike Powerful Owls, they live in most habitats including farmland, high plains, desert, rainforest, taiga forest, mountains and wetlands – so many habitats, in fact, that it is difficult to discriminate typical from atypical habitat. This differentiates them from Powerful and Rufous Owls that prefer woodland and forest.

Great Horned Owls give four or five deep resonant hoots in each calling bout (Australian *Ninox* give two). This soft dovelike baritone has a lovely timbre on cold nights and is used to add midnight ambience to feature films, even films set in Australia. Often, Great Horned Owls are associated with the other common, large North American raptor, the Red-tailed Hawk. As owls build no nest, Great Horned Owls nest in potholes on cliffs, in tree hollows, broken-off stumps and in crow, raven and Red-tailed Hawks' nests in copses of trees on the plains around Winnipeg. Sometimes they usurp nests by breeding earlier than the nest's previous owner. As with Australian *Ninox*, they lay and defend clutches of two or three white eggs from predators, including humans, though Australian *Ninox* prefer tree hollows and rarely use stick nests or cliff ledges.

Great Horned Owls can be dangerous. Anyone visiting a few nests will eventually be hit by an attacking owl. Arthur Cleveland Bent (1938), in *Life Histories of North American Birds of Prey*, said about pairs in Massachusetts that they were '... flying about near the intruder, alighting in nearby trees, snapping their bills, and hooting. Once I was savagely attacked, while I was climbing to a nest in which the eggs were hatching. I had hardly climbed 10 feet on the big pine tree, when the great brown bird glided past me and alighted in a pine beyond. There she sat glaring at me, swaying from side to side, her wings partly spread, her plumage ruffled out, looking as big as a bushel basket, her ears erect, and snapping her bill furiously, a perfect picture of savage rage ... Once, when I was not looking, I felt the swoop of powerful wings, then a terrific blow on my shoulder, almost knocking me out of the tree, and I could feel the sharp claws strike through my clothes. Several times I had to dodge from furious attacks. As I neared the nest, I felt a stunning blow behind my ear, which nearly dazed me, and off sailed my hat a hundred feet away; her sharp talons had struck into my scalp, making two ugly wounds, from which blood flowed freely. This was the limit; I did not care to be scalped, or knocked senselessly to the ground, so down I came, leaving the owls the masters of the situation' (p. 313).

Though somewhat anthropomorphic, Bent captures the ferocity of a nesting Great Horned Owl. Once, near Pullman Washington, researchers were banding nestling Great Horned Owls in an abandoned Red-tailed Hawk's nest. The adult female sailed straight at the photographer; he assumed she was bluffing. She wasn't. He focused his lens as she came in, wings fixed and silent, and she smacked him in the forehead, cut him and knocked him backwards. He was lucky; she could have struck the camera lens and cut his eye. Large Australian

Ninox also defend against humans, though, in contrast to Eagle Owls, the male is more aggressive than the female.

In spite of heavy persecution from shooting, trapping, poisoning, road casualties and more, Great Horned Owls remain common. How does such a dangerous bird remain abundant from Tierra del Fuego to the Arctic when wolves and other predators are shot, trapped and poisoned out? Partly, for the same reasons that Coyotes prevail; Great Horned Owls are secretive. Even in built-up areas, they are difficult to detect. And they are versatile. Eagle owls hunt less by listening for prey than many owls do, and they cannot find prey at extremely low light levels. They hunt primarily by perching at vantage points then making short flights of maybe 100 m to capture prey. The account of the Great Horned Owl in another classic text, *Birds of America*, edited in 1919 by scientists led by T. Gilbert Pearson (reprinted in 1936) begins: 'Tiger of the Air is the term which has been applied to this great Owl and fitly too, it must be admitted, for the big bird undeniably is courageous, powerful, and bloodthirsty. That he is highly destructive must be conceded, for it has been demonstrated beyond question of a doubt that not only that he is bold, persistent and generally successful in raids upon domestic poultry of all kinds, but that he is highly skilful and deadly in his pursuit of game birds, song birds, and squirrels. The tiger comparison applies well to the Owl's hunting, for the sweep of his great wings in the silent air is as noiseless as the tread of a cat's padded feet upon the soft earth. Through the woods and over meadows he glides as silently as a shadow of certain and sudden death. For such creatures the Owl's lightning-like swoop, and the murderous clutch of its great talons, are as pitiless and as inescapable as the spring of the tiger on a helpless lamb' (p. 113).

Their grip is powerful; it requires 13 000 g of force to open the talons once they are closed, and this allows them to tackle heavy prey. They take domestic cats, skunks, porcupines, woodchucks, rabbits, hares and many birds including geese, herons, pheasants and grouse, but they also take insects, scorpions and other small prey. They eat other species of owl, and find enough food in most habitats to build up fat reserves and breed earlier than most raptors do, in winter with snow on the ground at cold northern latitudes. You can see females incubating eggs in February with snow piled on their heads and backs. As the heaviest North American owl breeding south of the Arctic, they dominate other bird species, and this adds to their success. In the Yukon, they can live on Snowshoe Hares, or in Iowa farmland on small *Peromyscus* deer mice.

Table 14.1 shows the results of one diet study from northern North Dakota.

Some early bird-lovers hated Great Horned Owls – A.C. Bent again – 'I once had a captive horned owl that I raised from the nest and kept in my aviary with several other hawks and owls; in the next cage, separated by a chicken-wire partition, was a pet red-shouldered hawk, of which I was very fond, as it would eat out of my hand; one night the owl broke through the partition, killed and partially

Table 14.1 Percentage composition in prey used by Great Horned Owls (after Murphy 1997)

	Frequency		Average weight	Total biomass	
Prey category	n	%	grams	kilograms	%
Mammals					
White-tailed Jackrabbit	95	3.3	452.6	43.0	7.5
Ground squirrel	42	1.4	240.5	10.1	1.8
Mouse	385	13.3	16.6	6.4	1.1
Vole	328	11.3	29.5	9.7	1.7
Muskrat	20	0.7	475.0	9.5	1.7
Norway Rat	87	3.0	203.0	17.7	3.1
Miscellaneous	14	0.5	100.0	1.4	0.2
Subtotal		**33.5**			**17.1**
Birds					
Grebe	99	3.4	358.0	35.4	6.2
Duck	1010	34.8	253.9	256.5	45.0
Grouse and partridge	34	1.2	511.8	17.4	3.1
Rail	315	10.9	371.7	117.1	20.5
Shorebird	79	2.7	96.2	7.6	1.3
Passeriform	303	10.4	56.7	17.2	3.0
Domestic chicken	20	0.7	774.0	15.5	2.7
Miscellaneous	18	0.6	200.0	3.6	0.6
Subtotal		64.7			82.4
Amphibian	**38**	**1.3**	**50.0**	**1.9**	**0.4**
Reptile	**1**	**<0.1**	**100.0**	**0.1**	**<0.1**
Insect	**12**	**0.4**		**<0.1**	**<0.1**
Total	**2900**	**100.0**		**570.1**	**100.0**

devoured my pet hawk; the owl soon paid the penalty of a murderer and is now in my collection' (p. 309).

Modern Great Horned Owls pay similar penalties. Biologists and falconers reintroduce threatened species such as Peregrine Falcons into the wild; they do not list Great Horned Owls among their favourite birds. Great Horned Owls eat large falcons. Biologists try to build up collapsed or extinct populations by releasing (hacking) young Peregrines just learning to fly. Great Horned Owls eat these fledglings. In this process nestling raptors are locked in a 'hack' box, in a tree or on a cliff, and fed every day at the same place, then released just before they can fly. No adults are involved. The fledged young return to feed most days from the box, and they have a month or longer to practise hunting and gain independence. Great Horned Owls capture these naive young falcons, and owls catch more hacked falcons than parent-raised fledglings because hacked falcons have no protection. Falconers and researchers trap and remove Great Horned Owls from some hack areas, or occasionally shoot them.

To conclude, Great Horned Owls and Powerful Owls are about the same size and weight, but they differ. No large Australian owl is so abundant or widespread. No large Australian owl regularly takes raptors as prey, or shows such flexibility in prey choice and breeding habitat.

15

Great Grey Owls and wing formulae

Out of Vassar, Manitoba, in woodland stripped bare by winter, an owl the colour of fireplace soot rested on a broken-off limb. It looked bigger than a Powerful Owl. The temperature was –21°C with sharp flakes of snow blowing against my face. Through the glasses I could see a Great Grey Owl *Strix nebulosa*.

The generic name, *Strix*, comes from both Latin (*strix*) and Greek (*strixg*), and denotes a kind of owl, especially a strident one (screech-owl). The species name *nebulosa* comes from the Latin for 'clouded' in reference to the owl's plumage.

In dimensions, this owl is larger than the Great Horned Owl or Powerful Owl. It is sooty grey with irregular dusky stripes and lighter greyish-white eyebrows and lores (space between the eye and the bill), and light chin marks. Its beak is dull yellow, its eyes bright lemon-yellow. It has a prominent facial disc for funnelling sound into its asymmetrical ears, one set higher than the other on the side of its head so the owl can triangulate distances and positions of minute sounds in the snow of moving prey. Boobooks and other *Ninox* such as Powerful Owls do not have asymmetrical ears. They hunt more by sight than by hearing.

The Great Grey is the largest 'wood owl', the owls in the genus *Strix*. Others include the Spotted and Barred Owl in North America, and the Tawny Owl in Europe. By dimensions, it is the largest owl in North America, about 65 cm long from beak to tail, compared to 58 cm for the Snowy Owl and Great Horned Owl, and about 55 cm for the Powerful Owl.

A.C. Bent in 1938 said of this enormous owl, 'its long wings and tail, its large round head, and its long, fluffy plumage make it seem much larger than it really is.

I have skinned several of these owls and have always been impressed with the surprisingly small, and apparently weak, body in a great mass of feathers' (p. 213).

Compared to Great Horned, Snowy and Powerful Owls, the Great Grey is mostly feathers. But, in defence of its nest, it is not as weak as Bent reported. The bird was little known back in 1938, and Bent and his colleagues thought Great Grey Owls did not regularly breed south of Canada. Their stronghold is in Canada, but an arm of their breeding range extends south along the Cascade Mountains into California, and pairs in the mountains readily attack humans who climb to their nests, so vigorously in fact that the owls sometimes injure themselves.

It seems odd that such a massive owl chooses to live in dense forest. In Manitoba they prefer mature Tamarack-Black Spruce conifer woodland during the summer. Tamaracks are a conifer tree with needles, but the needles turn gold in late summer. With the sun behind them, a Tamarack glows like a candle. The needles drop in autumn, at the same time leaves drop off tall willows, Balsam Poplars and other deciduous trees. They stay naked through winter.

For such big owls, they live on comparatively small prey – voles, mice, gophers and other small mammals. Mikkola (1983) examined nearly 5200 prey items from the breeding season and 87.7 per cent of prey species weighed between 10 and 49.9 g. Only about 10 per cent averaged more than 100 g, much smaller on average than Great Horned Owl and Powerful Owl prey. Bull and Duncan (1993) calculated from 4647 prey items in four regional studies that *Microtus* voles comprised 30.8 to 83.5 per cent of the total items (see Figure 15.1). Winter (1986, 1987) found that the average weight of 662 prey items from California was about 75 g, with Pocket Gophers *Thomomys bottae* contributing about 57 per cent of prey items and 80 per cent of biomass.

Great Grey Owls target scurrying small mammals they can see and hear. They launch their attacks from perches, or from a silent hover over the snow, and sometimes punch through the surface crust. They even crash through the roofs of gopher burrows. These punch-holes with crescent wing marks in the snow on either side of them are signs that a Great Grey Owl lives in snow-covered forest nearby.

Great Grey Owls have very broad wings, but broad or narrow wings do not relate to the relative size of an owl's prey. Wing shape relates to habitat. Great Grey Owls use their broad wing tips to hover silently, and manoeuvre through dense forest. They breed in Tamarack-Black Spruce forested wetlands, often close to a meadow or open grassland; they avoid hunting in the middle of large open treeless areas. In areas where Tamaracks are too small and scrubby to hold a big raptor nest they nest in Black Ash and Basswood adjacent to bogs and swampy forests (Johnsgard 2002).

The shape of a bird's wing is partly determined by the shape of the primaries, the long, outside flight feathers. This relates to diet and how they hunt. Owls have

Figure 15.1 Canadian biologist Jim Duncan calculated that Great Grey Owl prey was comprised of 30.8 to 83.5 per cent *Microtus* voles.

10 primaries and a comparison of the longest of these flight feathers to the shortest is the 'wing formula', the shape of the wing tip in flight. It is a statement of the relative lengths of all the primaries, e.g. 8 > 9, 7 > 10, 6, 5 > 4 > 3 > 2 > 1.

Owls such as Northern Hawk-Owls have relatively pointed wings, but Great Grey Owls have more rounded wings. The 10 primaries (flight feathers) on an owl are counted 1 to 10 from the innermost to the outmost (towards the tip) primary. If the longest feather is towards the inner wing, the wing tends to be more rounded. If the longest feather is towards the tip of the wing, as in falcons, the wing is more pointed. With the Northern Hawk-Owl the 8th primary from the inside of the wing is the longest, similar to all six Australian falcons which have the 9th primary longest, though Brown Falcons sometimes have the 8th longest. Southern Boobooks and Rufous Owls have the 7th primary longest, making the wing more rounded than the wing of a Northern Hawk-Owl and similar to the wing of

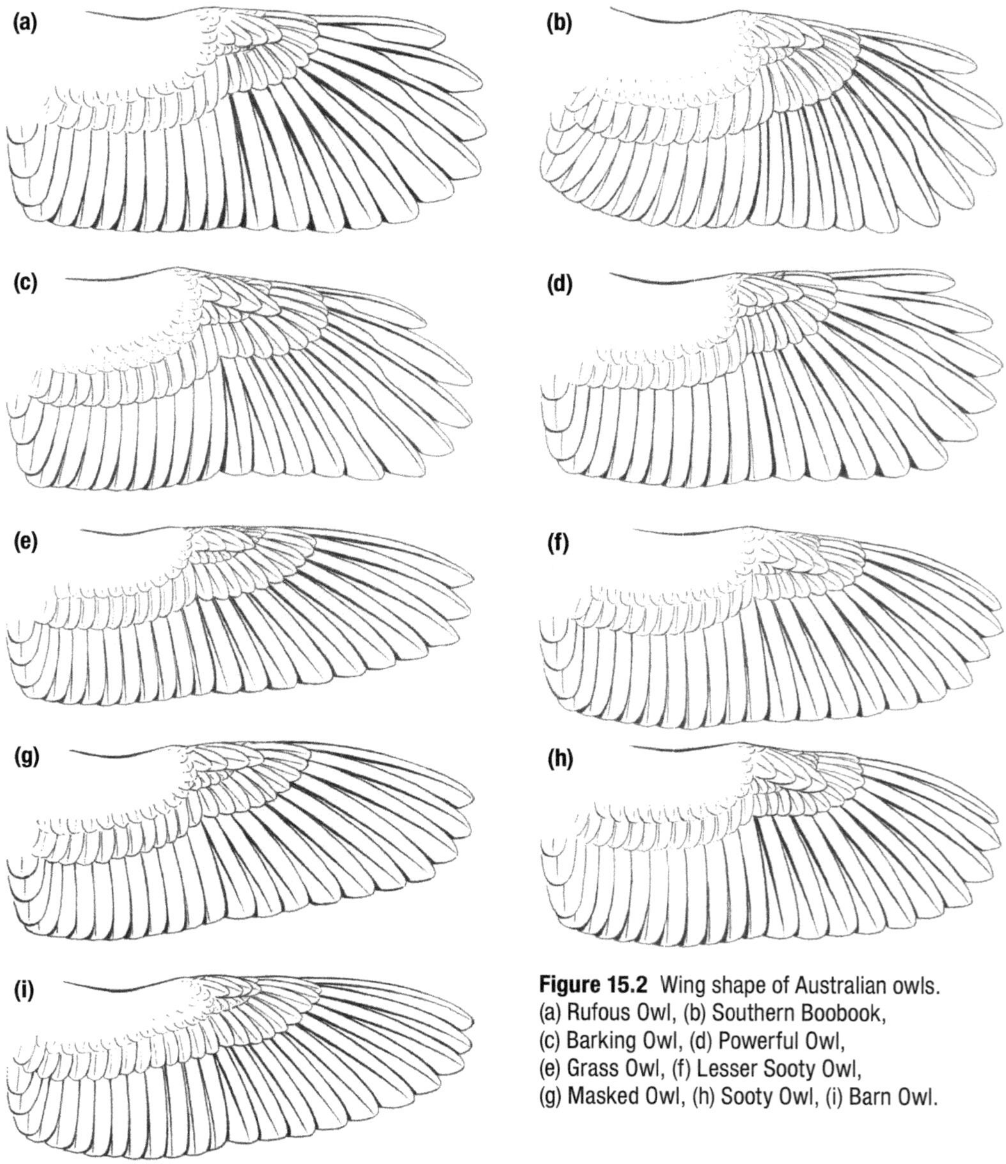

Figure 15.2 Wing shape of Australian owls. (a) Rufous Owl, (b) Southern Boobook, (c) Barking Owl, (d) Powerful Owl, (e) Grass Owl, (f) Lesser Sooty Owl, (g) Masked Owl, (h) Sooty Owl, (i) Barn Owl.

forest-hunting Australian goshawks and sparrowhawks which all have 7th the longest (wet forest dwelling Grey Goshawks sometimes have the 6th longest). Barking Owls have a slightly more pointed wing, the 7th or sometimes the 8th primary is longest. Great Grey Owls and Powerful Owls have the 6th primary longest, making their wing more rounded for flying and hunting in forest. Australian *Tyto* owls vary – Sooty Owls have the 8th longest with the 7th nearly as long, Lesser Sooty Owls have the 7th longest with the 8th nearly as long. This might look as though Sooty Owls hunt in more open forest than do Powerful Owls, but owls, like other birds, are confined by their phylogeny. This change in wing shape between species is limited by variation within the genus. Masked Owls have the 9th longest, and non-forests species such the Barn Owls and Grass Owls have the 9th or sometimes the 10th longest. (Figure 15.2 (a) to (i) shows wing shape for different Australian owls.) North America has no forest-dwelling *Tyto* owls and, if they did, the wing formulae would be different from that of forest-dwelling Great Grey Owls.

The Great Grey Owl on the end of the broken-off limb near Vassar sailed over the road like a lumbering bomber into naked grey timber too dense for such a big owl, or so it seemed. It folded its broad wings a little, flew through a small gap in the tangle of limbs and disappeared.

16
Powerful Owl diet

Owl specialists in Europe and North America often ask about Australian *Ninox* because of the Powerful Owl. The common name 'Powerful' and scientific names *Ninox strenua*, the 'strong' Hawk-Owl, summon up a half-hawk, half-owl hybrid, a yellow-eyed, broad-shouldered, forest dweller staring down on mortals. And often, that is how you find one.

Powerful Owls dwell in wet or dry eucalypt forests in south-eastern Australia, mainly on the east or coastal side of the Great Divide. Their range is limited by a line drawn from Rockhampton, Queensland, in the north down through New South Wales, around the belly of Australia to south-eastern South Australia. As forest birds they tend to avoid deserts and dry places. Some of them do, however, live inland, west of the Dividing Range at certain points – in Queensland as far west as Carnarvon Gorge, in New South Wales as far west as a north-south line joining Yass to Tenterfield, in Namadgi and Tidbinbilla in the Australian Capital Territory (Higgins 1999).

The closely related Southern Boobook takes spiders, insects, small birds and mammals, but Powerful Owls prefer larger mammals. They are not solely confined to tall open forest while hunting. They hunt where there is ample prey to support them for a time, then they move on. They can live in open woodland and cities, such as botanic gardens and forested suburbs in Sydney and Brisbane where they prey on birds, fruit bats and possums. Despite their large size, we know less about Powerful Owls than North American ornithologists know about large owls such as the Great Horned or Great Grey Owl, because Powerful Owls are more difficult to study. We know little about how they sneak through forest and ambush prey, but we do know which prey they ambush. As mentioned above, owls differ from most

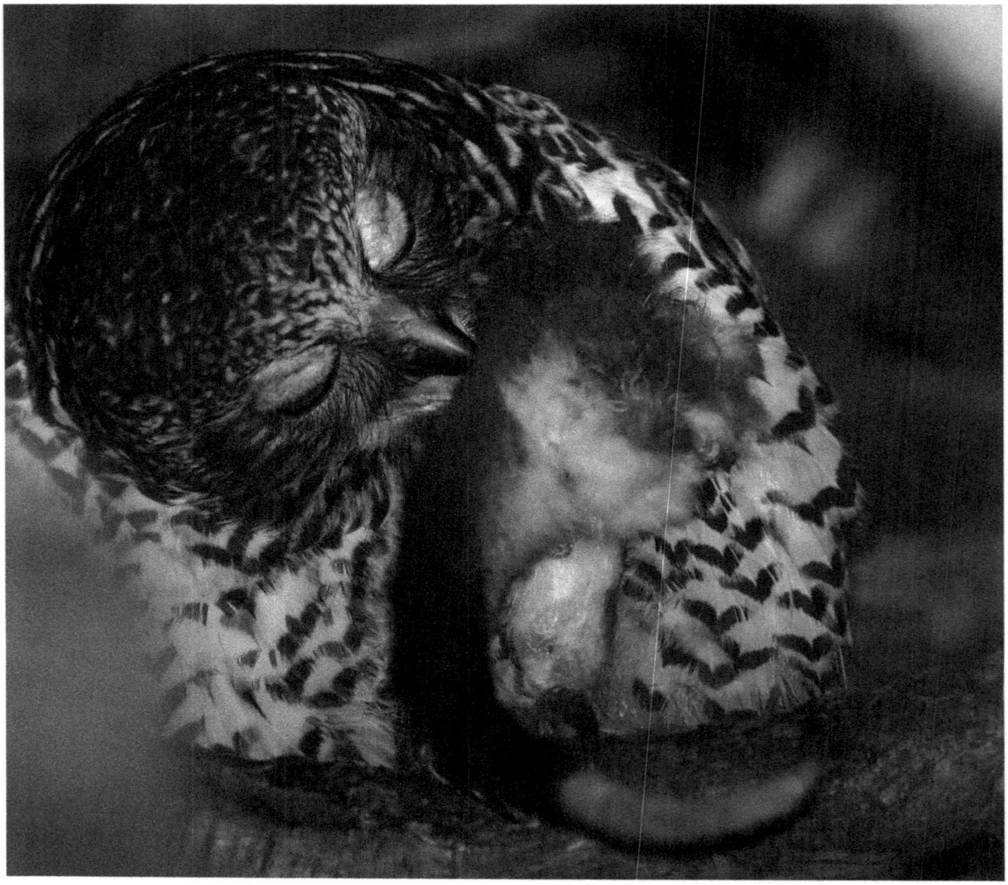

Figure 16.1 Powerful Owls prefer arboreal marsupials, especially Common Ringtail Possums.

other raptors like goshawks, which metabolise some bones they swallow. Goshawks have stomach acids 20 times as powerful as the acids in owls' stomachs so they can metabolise bone that owls cannot. This benefits owl biologists because the castings contain intact large and small bones, whole skulls, feet and fur that an owl has swallowed. With these clunky castings filled with whole leg bones, bits of fur, teeth and feathers, we can more easily identify prey species and count prey items.

Analysis of castings and prey remains show that Powerful Owls prefer medium-sized arboreal marsupials, especially Common Ringtail Possums and Greater Gliders (see Figure 16.1). They also take Sugar Gliders – the small marsupials that live in family groups, lick sap from tree trunks and look remarkably like the placental Northern Flying Squirrels taken by Spotted Owls (Figure 16.2). Powerful Owls occasionally come to the ground to catch Brown Hares or European Rabbits but, unlike Great Horned Owls, they prefer not to. They take few ground prey. They hunt the space above the ground, the tree canopy.

Figure 16.2 Powerful Owl day-roosting on a Sugar Glider in Canberra.

Powerful Owls dismember quite large prey, like Ringtail Possums and Greater Gliders, and the pelt, bones, feathers and tails can be collected from under roosts. Greater Gliders live almost exclusively on eucalypt leaves, as do Koalas whose young are said to be snatched off their mothers' backs by Powerful Owls. Like Koalas, Greater Gliders need a very large caecum, the blind end of the first part of the large intestine, to break down the cellulose in eucalypt leaves through bacterial fermentation. This is the caecum you see hanging from eucalypt and acacia branches in Powerful Owl country. Owls pull them out and discard them. Visiting an owl site in the forest you might see skins and tails on the ground and the caecum and intestines of gliders strung through tree branches. They discard these because they do not like the taste of gut, or maybe the contents of eucalypt leaves are toxic or distasteful and inedible.

They also hunt Black and Grey-headed Flying-foxes and birds such as Australian Magpies, White-winged Choughs, Pied Currawongs, Crimson Rosellas,

Table 16.1. Number of prey items of Powerful Owls by number and biomass from Australian National Botanic Gardens (Botanic Gardens) and Bendora Arboretum (Namadgi) (after Olsen *et al.* 2011)

Prey species	Botanic Gardens	Namadgi	Mass (g)	% by n	% biomass
Mammals					
Sugar Glider – adult	38	5	125	69.4	41.1
– sub-adult	2		65	3.2	1.0
– juvenile	1		50	1.6	0.4
Greater Glider		1	1300	1.6	9.9
Common Ringtail Possum – adult	1		846	1.6	6.4
– sub-adult	3		627	4.8	14.4
– juvenile	6		335	9.7	15.4
Common Brushtail Possum – sub-adult	2		570	3.2	8.7
Birds					
Crimson Rosella		1	131	1.6	1.0
Bassian Thrush		1	100	1.6	0.8
Invertebrates					
Crayfish		1	115	1.6	0.9
Total	**53**	**9**		**100.0**	**100.0**

and even beetles and other invertebrates.[1] They catch Brown Goshawks, Laughing Kookaburras, White-faced Herons, Sulphur-crested Cockatoos and Glossy Black-Cockatoos.[2] But how Powerful Owls do this remains a mystery because they are different from large owls in North America, they rarely or never hunt in daylight. Only a few people have ever seen them hunting at night.

In the winter of 2007 a single Powerful Owl turned up in the National Botanic Gardens on Black Mountain in Canberra. It was first spotted by Anthony Overs of the Canberra Ornithologists Group. The sex was never determined, but the owl sat in low trees, sometimes clutching prey. Bird-watchers lined up with binoculars and cameras. Tempers flared between crowds of bird-watchers and some Botanic Gardens staff until rangers roped off an area to give the owl space and protection from disturbance. This worked.

1 The following was taken from McNabb 1996; Higgins 1999; Debus and Chafer 1994; Kavanagh 2002a; Pavey *et al.* 1994.

2 From Tilley 1982; Lavazanian *et al.* 1994; Fleay 1968; Davey 1993; Higgins 1999.

Table 16.2. Minimum number of prey items taken by Powerful Owls in News South Wales (from Kavanagh 2002a)

Prey item	Total MNI	Weight (g)	Estimated biomass (g)
Aboreal mammals			
Common Ringtail Possum#			
– juvenile (61%)	591	335	197 985
– sub-adult (27%)	262	627	164 274
– adult (12%)	116	846	98 136
Greater Glider	193	1300	250 900
Yellow-bellied Glider	10	575	5750
Sugar Glider	114	125	14 250
Mountain Brushtail Possum (juvenile)*	1	570	570
Common Brushtail Possum (juveniles)#	29	570	16 530
Eastern Pygmy Possum	0	25	0
Koala	4	6000	24 000
Grey-headed Flying-fox	30	675	20 250
Microbat (*Nyctophilus?*)	0	14	0
Terrestrial mammals			
Red-necked Wallaby – juvenile	1	2500	2500
Long-nosed Bandicoot	1	850	850
Southern Brown Bandicoot	0	700	0
Brown Antechinus	0	25	0
Dusky Antechinus	0	50	0
White-footed Dunnart	0	23	0
Bush Rat	0	125	0
Black Rat	4	125	500
Swamp Rat	0	125	0
Broad-toothed Rat	0	125	0
House Mouse	0	13	0
European Rabbit	0	1500	0
Other			
Birds	223	350	78 050
Reptiles (Eastern Water Dragon?)	2	500	1000
Crustaceans (freshwater crayfish?)	2	115	230
Insects	89	0.5	44.5
Total (number of items)	**1672**		

#Common Ringtail and Common Brushtail weights based on Pavey *et al.* (1994).
*Estimate based on Common Brushtail.

This owl killed mainly Sugar Gliders (Table 16.1) but also Ringtail Possums and some baby Brushtail Possums, probably off their mothers' backs. The owl came to the Gardens for this abundant food and maybe used Black Mountain as a staging area before travelling to some remote gully in Namadgi to breed. Rangers Greg Flowers, Joe McAuliffe, Robyn Lawrence and Greg Sattler watched the owl after dark and took us along on some occasions. This owl was surprisingly buoyant; it fluttered like a big moth or a giant Boobook. It did not fly like the similar-sized Great Horned Owl that prefers more terrestrial prey (Table 14.1); it was more agile and could turn 180 degrees in mid-air. This made sense because Powerful Owls often catch prey in the tree-canopy and less often on the ground, and they need agility for this. The owl attacked Sugar Gliders in mid-air and thudded against tree trunks in misses. It ignored ground prey such as rabbits, abundant in the Gardens. The rangers said it attacked adult Brushtail Possums but it clacked its beak when it did so. It treated the adult Brushtails as competitors for hollows, not as prey. A small sample from a pair in Namadgi National Park took slightly different prey.

A large study by Kavanagh (2002a) showed more species but a similar pattern (Table 16.2).

Another study by Chris Pavey in Queensland showed a different pattern with more flying-foxes than arboreal marsupials (Table 16.3), but still this prey list underscored the Powerful Owl's preference for arboreal prey.

There are few data for the closely related Rufous Owl *Ninox rufa*, but large Australian *Ninox* take proportionally more arboreal mammals and birds than do similar-sized Great Horned Owls in North America. In Tables 16.1–16.3 you can see how Powerful Owls tend to ignore common ground prey such as rats, bandicoots and rabbits. This is not the case with all Australian forest owls. In the following chapter we compare the prey of two other forest owls to that taken by Powerful Owls.

Table 16.3 Minimum number of prey items taken by Powerful Owls in Queensland (from Pavey *et al.* 1994)

Species	Average weight (g)	Frequency	Estimated biomass (kg)
Mammals			
Common Ringtail Possum			
– pouch young	131	4	0.52
– juvenile	335	26	8.71
– sub-adult	627	13	8.15
– adult	846	6	5.08
Common Brushtail Possum			
– juvenile/sub-adult	570	13	7.41
Sugar and Squirrel Gliders[1]	121	18	2.18
Greater Glider[2]	1087	1	1.09
Black and Grey-headed Fruit-bats[1]	656	58	38.05
Little Red Fruit-bat[2]	358	1	0.36
Unidentified mammals	?	8	?
Birds			
Brown Goshawk (male)[3]	255	1	0.26
Brush Turkey (immature)[2]	800	1	0.80
Dusky Moorhen	234	4	0.94
Crested Pigeon[2]	182	1	0.18
Wonga Pigeon	380	2	0.76
Galah[2]	257	1	0.26
Rainbow Lorikeet	116	4	0.46
Scaly-breasted Lorikeet	80	20	1.60
Pale-headed Rosella[3]	145	1	0.15
Laughing Kookaburra[2]	306	1	0.31
Noisy Miner[2]	80	1	0.08
Olive-backed Oriole[2]	99	1	0.10
Grey Butcherbird[2]	75	1	0.08
Australian Magpie	301	2	0.60
Pied Currawong	321	2	0.64
Torresian Crow	538	9	4.84
Unidentified birds		7	
Insects			
Order Coleoptera Family Scarabaeidae[3]	2	1	*
Order Orthoptera Family Gryllacrididae[3]	2	1	*
Total		**209**	**83.61**

1. These two species are included together because it was not possible to consistently separate them on the basis of skeletal material, although both species were preyed upon.
2. Species recorded only as prey remains and/or prey held by owl(s).
3. Species recorded only in pellets.

*Weight not used in biomass calculations.

17
Forest owl diets

If large Australian *Ninox* prefer arboreal mammals and birds to ground mammals such as rats and rabbits, what about large *Tyto* owls in these same forests – the Sooty and Masked Owl?

The generic name for the Masked Owl, *Tyto*, is derived from the Greek for a type of owl, probably named for its cries. The species name *novaehollandiae* comes from the type locality New Holland = New South Wales. The distribution of Masked Owls is imperfectly known. They are said to occur over much of mainland Australia except the deserts. Wink *et al.* 2008 said the Tasmanian *castanops* was a separate species. Data for weights of these Tasmanian Masked Owls are limited but Fleay (1968) gave 595 g for a Tasmanian male and 1290 g for a Tasmanian female. Mainland Masked Owls are smaller and weigh less.

Masked Owls feed mostly on small- to medium-sized ground mammals, including rodents, but also arboreal or climbing mammals (Higgins 1999). In south-eastern New South Wales Kavanagh (2002a) found they took about 4.6 per cent arboreal mammals such as Sugar Gliders and Ringtail Possums, about 3.8 per cent birds, and about 90 per cent terrestrial mammals including Brown Antechinus *Antechinus stuartii*, Dusky Antechinus *Antechinus swainsonii* and Bush Rat *Rattus fuscipes* (see Table 17.1). On the central coast of New South Wales they took no arboreal mammals, about 70 per cent Black Rats (though Black Rats can climb, owls probably take most of them on the ground) and 20.5 per cent birds. Clearly the Masked Owl diet differs from the Powerful Owl diet and can differ from area to area, as Powerful Owl diet differs between new South Wales and Queensland (Tables 16.2 and 16.3).

Table 17.1 Prey list for Masked Owls from Kavanagh (2002a)

Prey item	% by number		MNI	MNI	Total number (MNI)
	SE NSW	Central Coast	SE NSW	Central Coast	
Aboreal mammals	**4.6**	**0**	**6**	**0**	**0**
Common Ringtail Possum	1.5	0	2	0	0
Greater Glider	0	0	0	0	0
Yellow-bellied Glider	0	0	0	0	0
Sugar Glider	3.1	0	4	0	0
Mountain Brushtail Possum	0	0	0	0	0
Common Brushtail Possum	0	0	0	0	0
Eastern Pygmy Possum	0	0	0	0	0
Koala	0	0	0	0	0
Grey-headed Flying-fox	0	0	0	0	0
Microbat (*Nyctophilus?)*	0	0	0	0	0
Terrestrial mammals	**89.4**	**79.5**	**117**	**35**	**152**
Red-necked Wallaby – juvenile	0	0	0	0	0
Long-nosed Bandicoot	0	0	0	0	0
Southern Brown Bandicoot	0	0	0	0	0
Brown Antechinus	29	0	38	0	38
Dusky Antechinus	22.9	0	30	0	30
White-footed Dunnart	0	0	0	0	0
Bush Rat	35.9	0	47	0	47
Black Rat	0.8	70.5	1	31	32
Swamp Rat	0	0	0	0	0
Broad-toothed Rat	0	0	0	0	0
House Mouse	0.8	9.1	1	4	5
European Rabbit	0	0	0	0	0
Other	**6.1**	**20.5**	**8**	**9**	**17**
Birds	3.8	20.5	5	9	14
Reptiles (Eastern Water Dragon?)	0	0	0	0	0
Crustaceans (freshwater crayfish?)	0	0	0	0	0
Insects	2.3	0	3	0	3
Total (number of items)			**131**	**44**	**175**

In contrast, the Sooty Owl diet appears to overlap more with both Powerful Owl diets, but they also take ground mammals and overlap with the Masked Owl diet (Bilney *et al.* 2006). The species name, *tenebricosa*, from the Latin *tenebricosa* for dark or gloomy, highlights the sooty-grey plumage (see Figure 17.1). Wink *et al.*

Figure 17.1 Rohan Bilney with trapped Sooty Owl.

(2008) considered Sooty Owls and Lesser Sooty Owls to be separate species. Weights for male Sooty Owls average about 650 g, females 990 to 1250 g, commonly over 1100 g according to Hollands (2008). Sooty Owls are said to range from mid-eastern Queensland (Eungella National Park) to southern Victoria. Lesser Sooty Owls occur in north-eastern Queensland between Cooktown and Townsville.

Sooty Owls prey mainly on arboreal mammals, mostly marsupials, including small possums and gliders, as well as terrestrial mammals, primarily rodents and small dasyurid marsupials, but also bandicoots, potoroos and rabbits. They sometimes take birds, lizards and large insects. Kavanagh (2002a) found that Sooty Owls in south-eastern New South Wales took roughly, by number, 48 per cent arboreal mammals including Common Ringtail Possums and Sugar Gliders, and 52 per cent terrestrial mammals including Brown Antechinus, Dusky Antechinus and Bush Rat. The remaining 1.2 per cent was mostly birds (Table 17.2). On the New South Wales central coast they took about 44 per cent arboreal mammals and

Table 17.2 Prey list for Sooty Owls from Kavanagh (2002a)

Prey item	% by number SE NSW	% by number Central Coast	MNI SE NSW	MNI Central Coast	Total number	Average weight (g)
Aboreal mammals	**47.6**	**44.3**	**502**	**182**	**684**	
Common Ringtail Possum	23.0	23.3	242	100	342	900
Greater Glider	0.4	2.9	4	12	16	1300
Yellow-bellied Glider	0.8	0	8	0	8	575
Sugar Glider	22.6	17.2	238	71	309	125
Mountain Brushtail Possum	0	0	0	0	0	4000
Common Brushtail Possum	0.2	0.2	2	1	3	3500
Eastern Pygmy Possum	0.6	0.5	6	2	8	25
Koala	0	0	0	0	0	6000
Grey-headed Flying-fox	0	0	0	0	0	675
Microbat (*Nyctophilus?*)	0	0.2	0	1	1	14
Terrestrial mammals	**51.5**	**48.1**	**542**	**198**	**740**	
Red-necked Wallaby – juvenile	0	0	0	0	0	
Long-nosed Bandicoot	4.6	3.9	48	16	64	850
Southern Brown Bandicoot	1.0	0	10	0	10	700
Brown Antechinus	10.6	10.9	112	15	127	25
Dusky Antechinus	8.2	0.2	86	1	87	50
White-footed Dunnart	0.2	0	2	0	2	23
Bush Rat	25.0	18.5	263	76	339	125
Black Rat	0	12.9	0	53	53	125
Swamp Rat	0.2	0	2	0	2	125
Broad-toothed Rat	0.1	0	1	0	1	125
House Mouse	0	0.7	0	29	29	13
European Rabbit	1.6	1.0	17	4	21	1500
Other	**1.2**	**7.5**	**12**	**31**	**43**	
Birds	1.0	2.2	10	9	19	350
Reptiles (Eastern Water Dragon?)	0.1	4.1	1	17	18	500
Crustaceans (freshwater crayfish?)	0	0.2	0	1	1	100
Insects	0.1	1.0	1	4	5	0.5
Total (number of items)			**1054**	**412**	**1466**	

about 48 per cent terrestrial mammals. As with Masked Owls in the central coast area, they took proportionally more Black Rats.

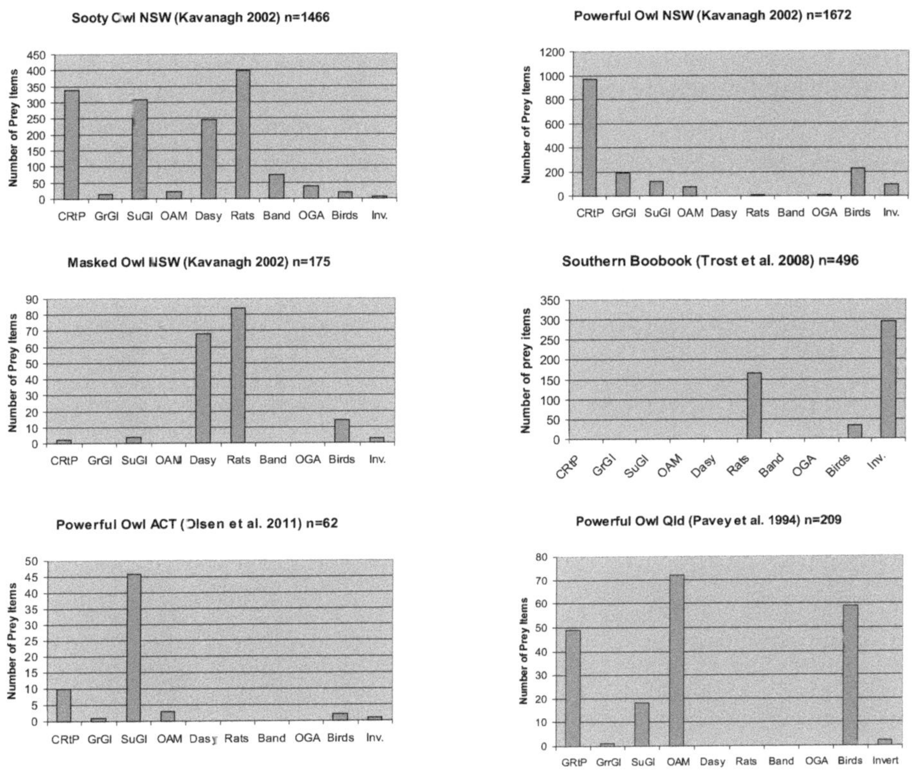

Figure 17.2 Proportions of Common Ringtail Possums (CRTP), Greater Glider (GrGl), Sugar Glider (SuGl), other arboreal mammal (OAM), Dasyurid (Dasy), Rats, bandicoots (Band), other ground mammals (OGM), Birds, and Invertebrates (Inv.) in diets of large forest owls and Southern Boobooks. Powerful Owls, Masked Owls, Sooty Owls in New South Wales from Kavanagh (2002a); Powerful Owls in Queensland from Pavey *et al.* (1994); Powerful Owls in the Australian Capital Territory from Olsen *et al.* (2011); Southern Boobooks in the Australian Capital Territory from Trost *et al.* (2008).

Comparing these large Australian owls with the North American Great Horned Owl, the latter is more ubiquitous and lives in a far greater range of habitats. It nests in forest and woodland as large Australian owls do, but also in desert and grasslands which large Australian owls do not do. Great Horned Owls hunt a wider variety of prey. The two large Australian *Tyto* take ground mammals as Great Horned Owls do, but are limited to forest habitat and a smaller range of prey.

The Sooty Owl diet is similar to the Powerful Owl diet in that both take arboreal mammals, especially Ringtail Possums, but it differs from the Powerful Owl diet in that Sooty Owls take proportionally more terrestrial mammals.

Looking at broad prey groups taken by Australian owls, there is similarity in the Powerful Owl samples collected by Kavanagh, Pavey *et al.*, and Olsen *et al.* (see Figure 17.2) but differences with Masked Owls and Sooty Owls.

As you can see in Figure 17.2, Powerful Owls take proportionally more arboreal prey, Masked and Sooty Owls take more terrestrial prey such as rats, and Southern Boobooks take no arboreal mammals. Instead, Boobooks concentrate on ground mammals, birds and invertebrates. In Chapter 18 we explore whether these patterns relate to dimorphism in these owls.

18

Powerful Owls and dimorphism

Wisps of light drift through gaps in the canopy. You catch an impression above you, a vertical form with striking dark chevron barring across a muscled front. The colour pattern, as if giving warning, is the opposite of the subtle browns of Southern Boobooks. Orange eyes stare down like light bulbs, as if annoyed (Figure 18.1). One foot is tucked up; the other grips a dead Greater Glider. High-arched black talons hold a bloody bundle draped over the branch. Other large owls, such as Great Horned Owls or Masked Owls, never do this. Why Powerful Owls?

As mentioned above, Powerful Owls exhibit NSD (Normal Sexual Dimorphism), with males larger than females. Most owls exhibit RSD (Reversed Sexual Dimorphism) – females larger than males. 'Reversed' is perhaps a sexist rendering of a principle in biology – males are larger and heavier because they compete for females, as in Peacocks or Jungle Fowl (domestic chickens), various species of cat and humans. Peacocks and Jungle Fowls have 'Normal' Sexual Dimorphism. In contrast, the Pygmy Owl, mentioned above, has females larger than males, as do Tawny Owls, Barn Owls, Eurasian Eagle Owls and Southern Boobooks. Great Horned Owls and Powerful Owls are about the same size but male Great Horned Owls weigh about 1300 g and females about 1700 g, compared with 1500 g for *male* Powerful Owls, and 1250 g for *female* Powerful Owls. Powerful Owls have switched, broken the rules.

Why have they evolved this way? Does the very long breeding cycle in Powerful Owls have anything to do with males being the heavier sex? Not likely, because many large eagles, such as Crowned Eagles in Africa or Harpy Eagles in South America, also have very long breeding cycles, even longer than that of Powerful

Figure 18.1 Powerful Owl.

Owls, and the female is the larger sex. The division of parental duties and the sex roles of Powerful Owls do not differ appreciably from the sex roles of other raptor species. Female Powerful Owls incubate eggs, and males do most of the hunting until nestlings are part grown; then females do more hunting.

To place Powerful Owls in a context with some other owls, Table 18.1 compares weights and degrees of dimorphism, how much males and females differ in size and in which direction. Most species in Table 18.1 were discussed above, though Tawny Owls and *Athene* species were not. They are discussed in Chapters 34 and 37. Tawny Owls are in the genus *Strix* with Spotted Owls and Great Grey Owls, discussed above. They specialise on mammals but take a variety of prey and breed across Eurasia in patches of woodland. *Athene* species are small owls with no ear tufts, not unlike little sandy-coloured Boobooks with longer legs. One species, the Burrowing Owl, nests in burrows dug by squirrels, prairie dogs, badgers and other animals on the open plains. Another, the Little Owl, like the Burrowing Owl, lives in open country and both are largely insectivorous.

In the ranking in Table 18.1, you can see that Great Grey Owls are the most dimorphic, and Burrowing and Little Owls the least dimorphic. Owls that hunt in forest, such as the Great Grey Owl, are often more dimorphic than owl species that hunt over open ground, such as Burrowing or Little Owls, probably because owls

Table 18.1. Approximate weight in grams of male and female owls discussed in the text and their rank using a simple Dimorphism Index (D.I.) dividing male weight by female weight

	Male	Female	D.I.
Great Grey Owl	900	1200	75.0
Great Horned Owl	1300	1700	76.5
Tawny Owl	470	580	81.0
Snowy Owl	1650	2000	82.5
Powerful Owl	1500	1250	–83.3*
Northern Pygmy Owl	63	73	86.3
Northern Hawk-Owl	300	345	87.0
Southern Boobook	280	320	87.5
Spotted Owl	570	650	87.7
Elf Owl	40	45	88.8
Morepork (SB in New Zealand)	177	193	92.0
Little Owl	172	175	98.3
Burrowing Owl	149	150	99.3

*Powerful Owl males are larger than females.

living in forest need agility to avoid collisions. Males do much of the hunting during the breeding season and their smaller size gives them more agility compared to females, as well as energetic efficiency when carrying prey over long distances. *Athene* species may be less dimorphic because they are insectivorous, so do not carry prey over long distances and live mostly in open country.

Powerful Owls catch large prey in comparison to their own body mass, and this may partially explain their habit of roosting during the day with prey under their talons (Figure 18.2). This sets Powerful Owls, Rufous Owls and Barking Owls (Figure 18.3) apart from non-*Ninox* owls. An eagle owl, like the Great Horned Owl of North and South America, would 'cache' large uneaten prey under a bush or hide it in a tree hollow instead of draping it over a limb in broad daylight where other predators could see it and take it. As mentioned above, a male Powerful Owl can weigh some 1500 g, the size of a domestic rooster, and females may be 1250 g. Common Ringtail Possums, a main Powerful Owl prey, weigh 700 to 1100 g. The Greater Glider, another principal prey species, weighs around 900 to 1700 g. That is, two of the preferred prey animals of Powerful Owls in certain parts of their range, the Greater Glider and Ringtail Possum, can weigh as much as the owl itself. Hunching at night on high branches in the forest, a Greater Glider looks the size of a house cat, except for its long fur and a long fluffy tail that stabilises it in long glides from tree to tree, or helps it careen into bushy cover to escape Powerful Owls. According to McKay (1983), they can glide 100 m, and switch direction by 90 degrees in mid-air. This suggests that Powerful Owls need agility to catch these gliders in forest and males should be smaller than females. But the opposite is true.

Figure 18.2 Powerful Owl with Common Ringtail Possum. It is not clear whether Powerful Owls significantly affect numbers of their prey.

We do not really know how Powerful Owls hunt prey, and maybe the large size of males gives them impact when they slam into large prey, and the power to kill, lift and dismember heavy prey. Maybe agility is of secondary importance.

With bird-catching falcons, for example Prairie Falcons in North America, there is said to be a '20 per cent rule' – the adult falcon tends to catch birds that weigh, on average, about 20 per cent of the falcon's body mass (Boyce 1985). This broadly holds with Peregrines. Powerful Owls, however, catch prey estimated by Kavanagh (2002a) to weigh 50 to 100 per cent of the owl's weight, and the predator/prey weight index in (Table 18.2) indicates that, using Geometric Mean Prey Weight, Powerful Owls can take prey that averages about 25 per cent of their own body weight. Some of the smaller Northern Hemisphere owls, such as the tiny starling-sized Pygmy Owl, also take large prey in comparison to their body size. They tackle woodpeckers or thrushes, Blackbird-sized birds, and sometimes ride them flapping along the ground in the snow for as long as an hour until they subdue them. This is a rare thing in raptors. Pygmy Owls take much of this large prey mainly during winter, but during the breeding season they take mainly smaller birds weighing no more than 35 g (Johnsgard 2002), small mammals and

Figure 18.3 Rod Kavanagh with a Barking Owl. Like Powerful Owls, Barking Owls sometimes roost holding prey during the day. Southern Boobooks do not do this.

some insects. Powerful Owls, and some other *Ninox*, are exceptional in the owl world because they take large mammals as their breeding diet. They are big, heavyweight owls that tackle big, heavyweight prey to feed their young. In the same way that we are attracted to big cats on the African plains that catch and struggle with large, dangerous prey in dramatic chases, we are drawn to Powerful Owls. As with these large African cats, you would expect Powerful Owls to catch juveniles if they could, because juveniles are smaller and less experienced at evading attack or fighting back.

In Colorado, Carl Marti (1974) found that Great Horned Owls took prey weighing 1 g to about 3000 g, averaging about 177 g, and higher than the mean 3 g for Burrowing Owls, 30 g for Long-eared Owls and 46 g for Barn Owls he found breeding in the same area. Marti *et al.* (1993) calculated Geometric Mean Prey Weight of prey, predator mass, and predator/prey mass ratio for six owl species breeding in Idaho – Great Horned, Barn, Long-eared, Western Screech, Burrowing and Northern Saw-whet Owl. In Table 18.2 these are compared to the same calculations for another study of Great Horned Owl (Murphy 1997), Great Grey

Table 18.2 Predator (source), Geometric mean prey weight, Predator mass, and Prey/predator ratio of owls in Australia and North America. Source: (1) Kavanagh 2002a, (2) Pavey *et al.* 1994, (3) Marti *et al.* 1993, (4) Murphy 1997, (5) Tishechkin 1997, (6) Olsen *et al.* 2006

Predator	GMPW (g)	Predator mass (g)	Prey/predator ratio
Powerful Owl (1)	386.7	1500	0.268
Powerful Owl (2)	323.22	1500	0.215
Great Horned Owl (3)	44.0	1500	0.029
Great Horned Owl (4)	43.41	1500	0.029
Great Grey Owl (5)	25.3	1030	0.025
Masked Owl (1)	74.66	590	0.127
Sooty Owl (1)	185.13	885	0.209
Barn Owl (3)	32.5	521	0.062
Northern Hawk-Owl (6)	24.9	295	0.084
Southern Boobook (6)	2.11	283	0.008
Long-eared Owl (3)	26.3	254	0.104
Burrowing Owl (3)	2.3	170	0.014
Western Screech Owl (3)	18.0	191	0.094
Northern Saw-whet Owl (3)	18.7	170	0.110

Owl (Tishechkin 1997), Sooty Owl, Masked Owl (Kavanagh 2002a), Powerful Owl (Pavey *et al.* 1994; Kavanagh 2002a), Northern Hawk-Owl and Southern Boobook (Olsen *et al.* 2006).

This may be a clue to the dimorphism puzzle. Powerful Owls catch large prey in proportion to the owl's body mass. There is some evidence with raptors that exhibit RSD that pairs with artificially increased broods or natural large broods respond by taking larger prey to feed more young instead of taking a larger number of the same size prey (Olsen and Tucker 2003 for Peregrines, Redpath *et al.* 2006 for harriers). No such experiments or studies exist with Powerful Owls, but maybe pairs with broods artificially increased from one or two to three would catch larger prey, as did Peregrines or harriers. Alternatively, they may take smaller prey because females catch more insects and common small prey around the nest to feed an artificially increased brood.

A second possible clue to NSD in Powerful Owls is that males are said to defend nestlings more vigorously against humans than females do, a marked difference from other birds of prey. Female Peregrine Falcons, Great Horned Owls, or goshawks defend the nest and young against human climbers more vigorously than males do, and all these species exhibit RSD.

A third possible clue to the mystery of RSD is the odd habit of perching with one foot on the body of last night's prey draped over the limb. They do not hide prey from other Powerful Owls. Large raptors such as Wedge-tailed Eagles and

Great Horned Owls do take large prey, but they never roost with it on a limb. The Eurasian Eagle Owl, much heavier and more powerful than a Powerful Owl, tends to hide surplus prey under a bush and retrieve it later. *Ninox* are alone in this perching-with-their-prey habit – they are the only owls that do so regularly. It may be a display, a way for males to control prey and give it in controlled amounts to females. Owl species with females larger than males, such as Great Horned Owls, cannot do this because the females simply bully food away from males. Closely related Southern Boobooks, with females larger than males, normally cache prey for the day and later retrieve it. Powerful Owls sometimes do this, but less often than Boobooks do.

Maybe Powerful Owls concentrate on large prey for two reasons: 1) in most environments small-bodied animals are more abundant than large-bodied animals; for example, mice are more abundant than deer or kangaroos (Newton 1998; Krebs 2008). But with arboreal marsupials, the opposite is true – larger possums and gliders tend to be more abundant than smaller ones, and more available as prey to Powerful Owls; 2) unlike Masked and Sooty Owls, Powerful Owls lack asymmetrical ears for triangulating sounds made by rodents and terrestrial mammals on the dark forest floor. So Powerful Owls specialise on larger arboreal marsupials in the canopy, while forest *Tyto* owls capture more small-bodied rodents and terrestrial mammals on the forest floor.

The size of prey taken by Powerful Owls, the male's role in nest defence and their habit of roosting on dead prey during the day may be clues to understanding 'Normal' Size Dimorphism in the enigmatic *Ninox*, Minerva's sacred Hawk-Owls.

Breeding

Compared to many birds, the reproductive period of owls is prolonged. Even the Southern Boobook, our smallest owl, takes at least five months to complete a breeding effort. Southern Boobooks breed seasonally, once a year, at least in the southern part of their range, in the spring. Most owls establish nesting territories large enough for breeding and large enough for both the male and female to hunt in. Territorial aggression early in the breeding season probably reflects this long-term goal. Owls show fidelity to these territories, reoccupying them in consecutive years. One or both members of the pair return annually to the same nesting area, often the same nest site, until death or some environmental change causes disruption to this pattern (Craighead and Craighead 1956). Such fidelity to one breeding area enhances breeding success and the longevity of the owls. They can learn all features of a hunting area, where prey lives, feeds and takes refuge, and the location of obstacles, predators and other dangers that cause mortality. Owls can also avoid overlapping the territories with other owls. One would expect the territories of insect specialists to be smaller than the territories of mammal or bird specialists.

19

Southern Boobook calls

At dusk, Australian owls slip away from their roosts and nests, hunting mainly at night. Because owls are so difficult to find and watch, researchers often listen instead. Understanding owl vocalisations is crucial to understanding owl breeding. Different voices identify different owls, the owl's 'intent', its links to allies, enemies, mates and family. When you read an account of behaviour and status (if the owl is common, vulnerable or rare), the author infers much of this information from listening to owls calling at night. Researchers in Australia see few of the owls they hear calling during surveys; they colour-mark or radio-tag comparatively few owls. Clusters of calling marked on a map denote a pair of owls. Many of these assumptions have never been tested. Some exceptions to this form of survey are the photographic work at nests by David Hollands, the Barking Owl research done by Rod Kavanagh and Matthew Stanton, and DNA studies by Fiona Hogan and Raylene Cooke.

Higgins (1999) noted that scientists had not studied in detail the calls of even common owl species such as Southern Boobooks. What do owl calls mean? What were they saying to each other? By watching colour-marked birds one can determine which owl made which call and how often they made the call. This may seem obvious, but researchers had operated on several deeply held assumptions, for example, that deeper voiced Boobooks were females, that two Boobooks calling at the same time were 'duetting', that Boobooks called more often when there was no moon, no cloud, less wind, less rain and so on. Were these assumptions true? And how did calling relate to territory size?

To study this, we walked into woodland at sundown, several nights each week if possible, stood under the nest or daytime roost of one of the owl pairs, and watched them from just before they left their roost or nest for part or all of the night. At dusk the top of a Boobook's brown head shows in the hollow, then white eyebrows, yellow-green eyes, and a finally a full-face. They move up and sit in the entrance, watching and listening, testing for their main tormentors – currawongs, cockatoos, and honeyeaters. When it is safe, they fly out to a branch.

We carried a radio-receiver and antenna to triangulate the owls' location, find each one and positively identify it by shining a torch on its legs to read the colour-band.

Seven calls

Southern Boobooks gave seven main calls (Olsen *et al.* 2002a): (1) Boobook calls, (2) Single Hoot, (3) Por call (croak), (4) Mating Squeal, (5) Bray, (6) Purr and (7) Trill. All seven have their counterparts in other *Ninox*, such as Powerful Owls or Rufous Owls, and non-Australian *Ninox* such as the Sumba Hawk-Owl.

Boobook calls

Owls give these calls with their beak closed (or so it appeared through binoculars), and chin feathers fluffed (Figure 19.1). With each 'boo-book' couplet they squeezed their wings twice against the body, as if using their body as an accordion to force out each 'boo-book' syllable. This call sounds something like a European Cuckoo, the two-note call that German clock-makers put into wooden clocks. Boobook calls have two distinct types, each used in different circumstances, though functions overlapped.

1) **Territorial Boobook calls** – these are louder calls, usually given from a high perch across the territory, or blasted into a neighbour's territory, sometimes in bouts as long as an hour or more.
2) **Contact Boobook calls** – mated pairs usually gave these calls to each other, often facing one another and usually in short bouts of one to three calls, often with shows of affection. These calls are more 'personal' or 'intimate', often part of mating behaviours like mutual face preening.

Territorial Boobook calls

Territorial Boobook calling is territorial singing, the same behaviour that songbirds use to space themselves out across habitats or challenge intruders. Blackbirds announce possession of a territory or sing to attract a mate, as do Boobooks. As with songbirds, Boobook singing was seasonal. The frequency of calling increased in late winter and spring, fell away in late summer and autumn

Figure 19.1 Southern Boobook giving territorial Boobook calls. The beak is closed and chin feathers fluffed out.

and disappeared in early to mid-winter (Figure 19.2). Females gave Territorial Boobook calls about a third as often as males did (26 per cent versus 74 per cent of observation nights for males). Females called mainly when they were alone near the nest, and before egg-laying, when they were new on a territory or after nesting failure when females moved to a new location inside the territory and tried again. They rarely called with their mates, but males often called on their own, or in 'Duels' with neighbouring males. Females called more often than males did in late winter and spring and engaged in territorial disputes, sometimes against neighbouring males (Figure 21.1).

In birds, territorial singing is usually confined to males. However, female Boobooks sat on exposed branches, alone and singing, before egg-laying – a time when males should guard them from other roving males. We did not believe females were calling to their mates because their mates didn't answer from nearby

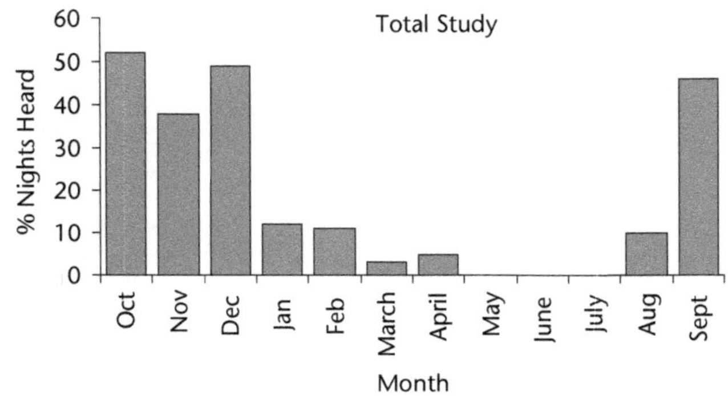

Figure 19.2 Per cent observation nights per month of 529 observation nights we heard Territorial Boobook calls (Total nights heard = 149/529 observation nights (28.1%)).

or from across the territory (though in a later study a few did this), and these females engaged in territorial disputes against other females, sometimes against neighbouring males. Why would a female defend her mate's border against another male? Some owl biologists believed these females were courting neighbouring males. This thinking stemmed from a narrow view about the roles of male and female raptors – males defend territory borders, females stay home and take care of babies. This generally is true; males and females do split breeding roles this way, but not all the time. It isn't that simple.

Why did females sing? Maybe their mates heard this territorial calling and 'knew' it was territorial defence, not a sexual display. Maybe this signalled to males that she was defending territorial borders, especially against other females, not courting other males.

Some writers had sexed Territorial Boobook callers on the assumption that males, being smaller, gave higher pitched calls than females. Bigger females had deeper voices. But in other owls, like Tawny Owls, Spotted Owls or Great Horned Owls, the smaller males gave deeper calls. This makes sense because males hold the territory longer, over-winter there and compete with each other for females. But we could not reliably discriminate male callers from females and we doubt that other researchers can without sonograms.

Contact Boobook calls

Contact Boobook calls don't carry as far as Territorial Boobook calls. They are muffled and lack timbre. These calls followed a similar seasonal pattern, but they started and finished earlier in the year. They peaked during pre-breeding and nesting (Figure 19.3). When we recorded occasional Contact Boobook calls as early as July, we assumed that a female had returned from her winter territory to make contact with her mate. Females were maintaining the pair bond, engaging

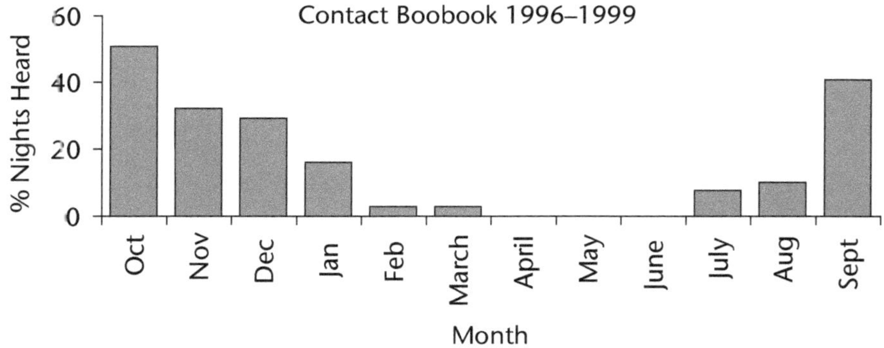

Figure 19.3 Per cent observation nights per month we heard Contact Boobook calls (total nights heard = 124/529 observation nights 23.4%; all male except 4/9/99 both male and female called; 25/9/99 female only; 8/10/99 male and female called).

in pre-breeding behaviour or keeping an eye on their mates, or females new to the territory were establishing bonds with a new mate. The pair would face each other; one would call gently, sometimes preening its mate's face, like African Lovebirds.

Males often gave this call when they arrived at the nest with food. The call was further muffled by a mouth full of fur, feathers or insects. Just before the male arrived, he often transferred prey from his feet to his beak, so he could deliver it beak-to-beak to the female. If he could not find the female, he did not eat the prey or give it to his young. He kept the prey and looked for her, all the while Contact Boobook calling. The Contact Boobook call and the food he brought were connected with pair bonding, like an offering, a gift.

Single Hoot calls

The Single Hoot is an alarm call that sounds like a single toot on a child's recorder. Adults used it when they defended the nest or young. It is like half a Boobook call but given with emphasis, exclamation. Boobooks used it most commonly after their young fledged, so in our study we heard it more often in December and January (Figure 19.4). Females used it just after their young fledged, sometimes as they swooped at our heads as if warning us off. Then males used it when they took over parenting duties, the guarding and feeding of young after females 'turned off' or left (see Chapter 30 on female desertion). A strange thing about Single Hoot calls is that Boobooks directed them only at humans, or perhaps at ground predators such as feral cats. They never used this call against possums, ravens or other enemies in trees or flying. Instead, they attacked possums or Sugar Gliders making a loud beak-popping noise as they swooped or hit them, never the Single Hoot call.

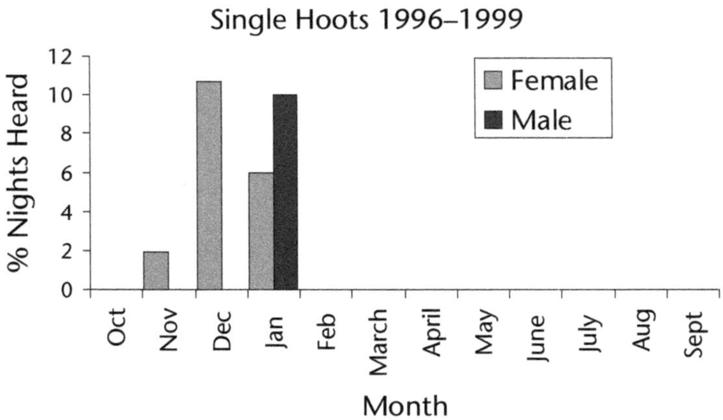

Figure 19.4 Per cent observation nights per month (18 of 529 observation nights) we heard Single Hoot calls; male 6 nights; female 12 nights.

Por (Croak) calls

Sometimes, at the beginning of the night or before a bout of Territorial Boobook calls, or when an owl met up with its mate or an enemy neighbour, they gave a series of deep, croaky notes. Other writers called them *more*, *por*, or *croaking* calls. These differed from Territorial Boobook calls that were two-note (disyllabic) segments. Por calls were a single low note repeated, rhythmically and monotonously, sometimes for long periods. The owls also made deep guttural Boobook calls, two-note calls. For this study we counted Croak calls and deep guttural (two-note) calls as the same, though later studies probably should split them. As with Boobook calls, these calls were seasonal (Figure 19.5).

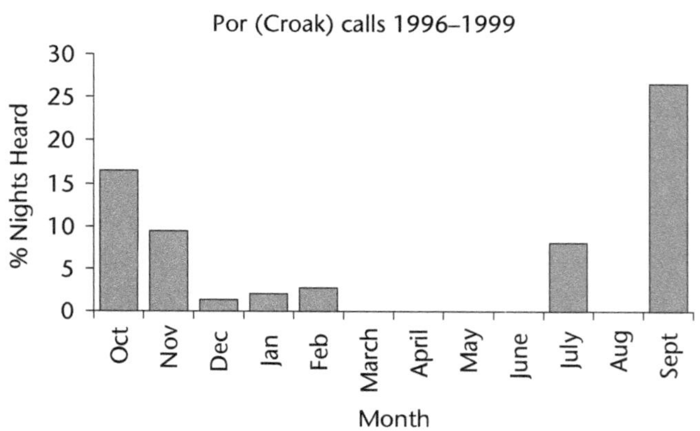

Figure 19.5 Per cent observation nights per month (39 of 529 observation nights) we heard Por (Croak) Boobook calls.

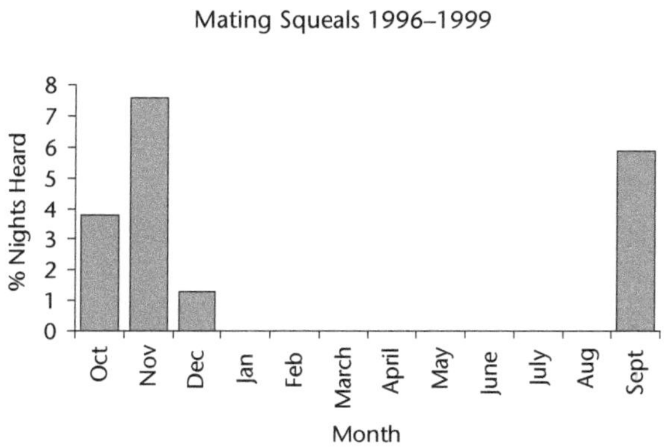

Figure 19.6 Per cent observation nights per month (12 of 529 observation nights) we heard Mating Squeals (all, we believed, given by females).

Owl language is still a mystery. Por calls have a sort of menacing quality, intimidating to a rival, but they can be sexual when used with a mate, when choosing a nest hollow or before copulation. This suggested that sexual and aggressive behaviours are close in Boobooks, maybe linked.

Mating Squeal

The Mating Squeal seems like an owl orgasm. It sounds like a rabbit caught by a fox or in a trap, a long quavering whistle. Only females seemed to give the Mating Squeal, and only after copulation. The comparison with orgasm was inescapable, but we hadn't seen a good explanation for the purpose of this call, or why females, not males, gave it. Mating Squeals lasted two to three seconds.

The distribution of Mating Squeals over the year (shown in Figure 19.6) probably did not represent normal frequencies because one pair in 1997 failed twice when neighbouring owls disturbed them. They tried three times to breed in three different nest hollows, and they copulated before each breeding attempt from September to December and gave Mating Squeals each time.

Bray

Females gave an open-mouthed Bray call (Figure 19.7) that sounded something like a Wood Duck, or a baby magpie calling. It is a food-begging call or a call to the male for assistance if predators are threatening the nest. Males responded to this call by arriving to fight possums or other threats, or by catching moths, other insects or sometimes catching a bird and giving it beak-to-beak to the female. As you can see in Figure 19.8, females started these food-begging calls before egg-laying and continued them through incubation, the nestling stage and the fledging of young. As in other female raptors, female Boobooks solicit food from males.

Figure 19.7 Female Southern Boobooks give Bray calls to solicit food or assistance from males.

The Bray call is used from the months when she is forming eggs to the months when the young fledge, maybe to induce males to hunt for their family.

Bray calls and Contact Boobook calls are secrets to finding Boobook nests during pre-breeding or incubation phases, before juveniles start Trilling. Territorial Boobook calls, the song most people associate with Southern Boobooks, are less helpful (Chapter 6). Territorial calls can be given at the nest, or up to 2 km from the nest. But Boobooks give Bray and Contact calls *at* the nest.

Purr

Some writers said that male Boobooks also gave the Bray (food-begging) call, higher pitched with thinner resonance than with females, but we never heard males give Bray calls. Males gave a softer Purr call, particularly to their young, as if communicating some kind of reassurance, especially after the young were flushed from a tree or otherwise disturbed. Both males and females gave this softer Purr call. We wondered if observers had confused the Bray call with this Purr call.

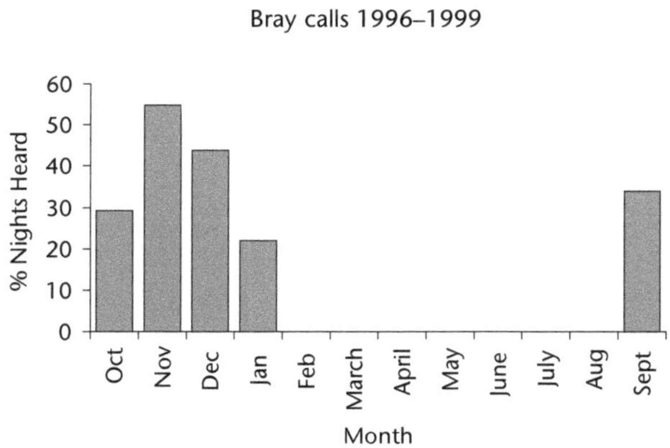

Figure 19.8 Per cent observation nights per month (119 of 529 observation nights) we heard Bray calls (all, we believed, given by females).

Trill

Nestling and fledgling Boobooks use the Trill call to food-beg and maintain contact with adults and siblings. It sounds like a cricket, or small frog, or a child blowing a tiny whistle. Fledged juveniles use it for much of the night, frequently while they fly from perch to perch. Often the adults filled up the juveniles with insects, bats or birds by the end of the first hour, then fledglings went quiet for a time.

On three occasions we saw fledglings cross territorial borders, Trilling as they did so. Neighbouring adults, the territory owners and enemies of these juveniles' parents, never fed them, even though juveniles trilled and begged for food from these neighbours. Neighbouring adults seemed to recognise that these border-crossing juveniles belonged to someone else and were no threat. They were never attacked. Many raptors, such as falcons, eagles and hawks cannot separate their own young from the same-species young of other pairs, and feed strange young along with their own. Maybe Boobooks recognised the Trill calls of both their own young and neighbouring young. Or so we thought at the time. It was not that simple, and we will examine this question further in Chapter 31.

There is talk in the forest between neighbouring owls, mated pairs and their young – warnings, intimidation, reassurance, alarm calls, signals. Different calls show how different owls are connected, who is hungry, who is an enemy and who is family. Calls help maintain stability.

20
Borders

Though breeding Southern Boobooks on Black Mountain fought with possums, gliders and other enemies (Chapter 11), they spent more time watching each other. When not hunting or caring for their families, they watched territory boundaries – the imaginary walls set through bushland. They patrolled these borders in case a neighbour strayed over, and they treated borders almost like stone fences. The intrusion by another owl could trigger a bout of Boobook calls, even a physical fight. The crossover could be 'unintentional', if intent can be attributed to an owl. Maybe the intruder failed to see the borderline, the line where his land stopped and his rival's land started. Or maybe it was 'intentional', that is, he was testing his neighbour's resolve in some form of contest. Most of this patrolling and conflict was limited to males, and mostly they seemed to know precisely where the borderline fell.

Sometimes a male would fly up to a border, stop and call to his neighbours. He would lean forward, call with the two-note Boobook call, repeating it for as long as an hour. If a Boobook responded, a contest might begin – owls yelling at each other on territory borders, sometimes across a gully or a road. Sometimes they called every night for weeks until the persevering intruder seemed to shift the border a few metres into his neighbour's patch. The male won a clump of forest or maybe a shaded gully, a grove of trees that benefited his breeding attempt. The loss of a good patch of forest probably hurt the defeated male.

On nights when conflict broke out, the males barely noticed human observers standing under the tree writing. They ignored Swamp Wallabies thudding past snapping sticks, or kangaroos bounding through crackling leaf litter.

To understand this territorial behaviour we colour-marked two adult males and two adult females (White and Green).

We defined a 'female' as the larger of the pair (by weight) with a brood patch. The brood patch was a naked patch of reddened skin on the lower abdomen, rich in blood vessels and used to keep eggs and nestlings warm on cold nights. Males never incubated, and never had a brood patch.

The sampling protocol was 30 minutes after sundown when it was just getting dark and the owls' heads just emerged, for 60 to 90 minutes after the first appearance, every night possible. This period covered the stages of breeding from egg-laying and incubation, through the period when young left the nest hollow, to late summer when juveniles left the care of their parents and the territory. We marked the location of owls on a map, the position of their day roost in a tree hollow, a leafy eucalypt crown or in a Native Cherry, and noted when each owl moved to territory borders and called.

Territories

The male Southern Boobooks in particular defended territory borders, in the same manner that many owls do. They sat in a tree, sometimes low in the foliage, sometimes high on an exposed branch, in woodland or in a suburban yard and gave the Boobook call. Each territory covered about 1 km^2, depending on the type of analysis used, and consisted of Canberra Nature Park forest and private gardens and yards. To stop another owl from stealing his ground, each male spent part of the night in spring and summer patrolling his scrap of land. The shape of a territory was not necessarily square, it could be long and narrow. When he had nestlings, male White often roosted during the day several hundred metres from his own nest, closer to the nest of the neighbouring male (Green) than to his own nest and young. This baffled us; we spent several nights following him. Maybe he had another nest somewhere, maybe he was feeding another female. But he was simply guarding his border and staying away from his nest and young. The neighbouring male, Green, had been pushing into White's territory and White meant to stop him. In Figure 20.1 you can see how Males White and Green were calling in an overlapping area in dispute. Table 20.1 shows how all four owls called inside core areas (where 50 per cent of activities were located) inside the home range (where 95 per cent of activity was located), but called sometimes on borders or outside of their home ranges.

It is important to understand that territory and home range are temporal concepts, not solely spatial concepts. Powerful Owls need enough space with enough prey in this space for them to rotate from species to species through the year as different prey becomes available, maybe in different parts of the territory. Boobooks apparently do the same. Prey abundance changes through the year, but

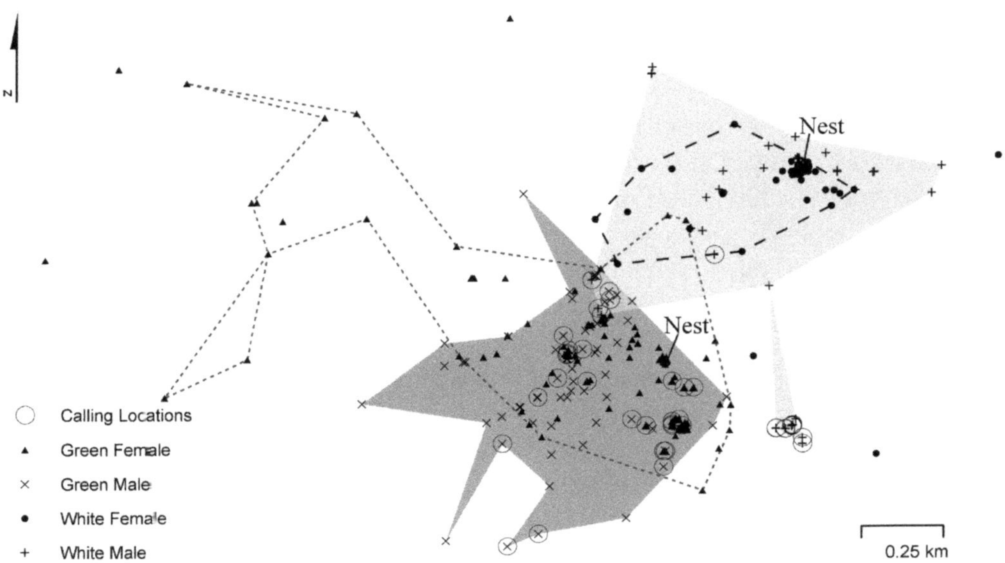

Figure 20.1 Home ranges of four Southern Boobooks during one breeding season (1998–1999) estimated using 95% characteristic hull polygons. Female home ranges are dotted or dashed, while males are solid. Locations of territorial calling are circled (after Olsen *et al.* 2011).

territory and home range stay fairly constant through territorial behaviour, especially calling.

There are several published accounts about territoriality and calling behaviour in Southern Boobooks, including: Fleay (1968), Schodde and Mason (1980), Hollands (1991), Shields (1994) and Debus (1996). Home range and a territory are different. A *territory* is defended, while a *home range* is not. Some claimed that Southern Boobooks were not necessarily territorial; they did not defend a particular patch; and home ranges were about 4 to 8 ha. But none of these authors

Table 20.1 Percentages of bouts of Territorial calling of Southern Boobooks that occurred within their home ranges (excluding core areas), and on the border or outside of their home ranges (after Olsen *et al.* 2011)

				Territorial calls (%)		
Owl	**Sex**	**Season**	**n**	**Core area**	**Home range**	**Border/outside**
White	M	94–95	5	60.0	40.0	20.0
White	M	97–98	5	0.0	60.0	40.0
White	M	98–99	16	37.5	25.0	37.5
White	F	99–00	10	10.0	70.0	20.0
Green	M	97–98	54	20.3	63.0	16.7
Green	M	98–99	19	42.1	47.3	10.5
Green	F	98–99	14	14.3	85.7	0.0

had colour-marked owls they studied, and researchers could not reliably identify different Southern Boobooks, even members of a mated pair, unless they were colour marked. Females are only about 10 per cent heavier than males, but this varies, and both sexes change their size and shape in the night by fluffing up if they were cold or angry, sleeking down when hunting, or crouching or stretching their necks, all of which made males look bigger or females look smaller.

The Boobook pairs behaved differently from Southern Boobooks in the literature. Was this due to our small sample size, or was the literature wrong?

The claim that Southern Boobooks were not territorial seemed to be a misreading of David Fleay's 1968 book. Fleay had said that, 'territory consciousness – so very pronounced in Powerful Owls and well developed to an extent also in the Barking Owl, is not so obvious in Boobook Owls – meaning that pairs may nest comparatively close to one another, as on our 65-acre Queensland Fauna reserve where three lots share this comparatively small area of bushland' (p. 94). Fleay's comment about lack of 'territory consciousness' being 'not so obvious' was taken out of context. He was referring to captive Southern Boobooks, because two females in the same enclosure had laid eggs and seemed to cooperate. But captive behaviour often differs from wild behaviour.

And Fleay's claim that three pairs in the wild shared 65 acres on his Reserve? Researchers may have decided to turn this statement into a study of home range or territory size, using simple maths to calculate home range size – 65 acres is 26.3 hectares divided by three pairs of Boobooks = 8.8 hectares. But Fleay never said this.

The closely related New Zealand Boobook (Morepork) does have smaller territory sizes, and they have been radio-tagged and studied, but this is a smaller owl than the Australian Southern Boobook. More importantly, the avifauna of New Zealand differs markedly from the avifauna in Australia. The Morepork has few competitors for food except other Moreporks. Southern Boobooks in Canberra caught birds, insects and mammals alongside Tawny Frogmouths, Australian Hobbies, Collared Sparrowhawks, Brown Goshawks, Peregrine Falcons, Little Eagles, Pied and Grey Currawongs, Australian Ravens, Laughing Kookaburras and Australian Magpies – all species that competed with Southern Boobooks for insect, bird or mammalian prey. Also, Canberra Boobooks competed for hollows with Laughing Kookaburras, Wood Ducks, Pacific Black Ducks, Brushtail and Ringtail Possums, Sugar Gliders, Galahs, Gang-gangs and especially Sulphur-crested Cockatoos. Little competition exists for Morepork hollows in New Zealand, except maybe from introduced Brushtail Possums, and Moreporks could use smaller nest hollows than those used by possums. Because of a lack of competitors, New Zealand Boobooks could have smaller home ranges and still harvest enough prey to breed.

There was, perhaps, another reason why three pairs of Boobooks could have appeared to live in such a small area on Fleay's Reserve in Queensland. Southern Boobooks did not necessarily nest in territory centres. They often nested on territory borders, close to another nesting pair. So three Boobooks nesting close together, as Fleay had seen, masked the fact that a large 100-ha territory could have stretched out behind each of these nests, or he could have heard them on his property but they ranged far beyond it. The nests were close, but the territories were large. This proved to be the case with Boobooks on Black Mountain.

21

Duetting and duelling

Different claims about the purpose of owl calling led to different interpretations of owl behaviour. The claim that mated male and female Boobooks 'duetted', sang together in a coordinated fashion was interesting and testable.

One can start with a simple question – 'Do Boobooks duet?' and then break this down into smaller questions:

1) How do ornithologists define duetting? (the first step in testing a claim)
2) How do we find evidence in the field showing duetting?, and
3) How valid is this previous evidence?

Definition

Marks *et al.* (1999) said about duetting Spotted Eagle Owls that mated males and females sounded like one owl calling. The two owls' calls blend into one, as happens with Australian Magpies or other butcherbirds. In her definition of duetting, Farabaugh (1982) first defined the term *bout*, crucial to her definition of duetting. She said: 'Vocalisations, like other behaviours, are clumped rather than randomly distributed in time. These clumps are called *bouts*' (p. 87).

She defined *duetting* as when 'bouts of certain elements in the repertoire of one bird frequently overlap with bouts of certain elements in the repertoire of its mate … Further, there is some organisation of both participants' elements within the region of overlap. This view can be expressed in terms of three variables that can be measured for any species: one measure of *bout overlap*, i.e. the percentage of bouts that overlap with bouts of the mate; and two measures of organisation of male and female elements within overlapping bouts, i.e. the *precision of timing* and the *sequential ordering of elements* ... The percentage of male bouts that overlap with female bouts, and vice versa, can be calculated for each type of bout (bouts of song, bout of each call type, etc.). If the percentage overlap is high, these overlapping bouts may be duets' (p. 87).

What exactly does this mean? First, that a male and female had to sing at the same time, their bouts of singing overlapped. Second, there was *precision of timing* and *coordination* in the calling. The calling was so tightly coordinated that it sounded much like one bird calling. Third, there was *sequential ordering* of the elements in the calling; for example, the order in a bout could be male-female-female-female, or maybe male-male-female-male-male-female, to make it sound like one complex song. The contribution from male and female was stereotypical.

With a definition in place, how can one look for evidence of duetting in the field? We concentrated only on the Territorial calls of Southern Boobooks – their territorial singing.

Bouts

Over 529 nights we counted the number of bouts, as defined by Farabaugh, and identified the callers. We counted 67 bouts for females, and 188 bouts for males. Sometimes two or three owls sat close together and we could not tell who the caller was unless we sighted colour-bands. By using the radio-receiver, we moved in close enough to sight one of the owls, then shone the torch on its legs. Usually we could not see the band from the front, so we moved around to the side or behind the owl and peered up under its tail.

We heard at least one bout of Territorial Boobook calling on 275 nights out of 529, about one every other night. On 74 per cent of the nights at least one caller was male, and on 26 per cent of nights at least one caller was female (Olsen *et al.* 2002b).

As mentioned above, the Territorial Boobook calls were seasonal (see Figure 21.1). They peaked during spring and dropped to zero during winter.

As you can see in Figure 21.1, both males and females called at similar times of the year, but females called less often, especially in late spring when they were incubating eggs and brooding nestlings, and they called less than males did in autumn. The females tended to call more often than males did just before they laid eggs and settled down to life as a parent.

Overlapping bouts

From Farabaugh's (1982) definition, we calculated the percentage of bout *overlap* between mated males and females, the times when mated pairs called together. As it turned out, only two of 255 bouts overlapped, 0.8 per cent, one on 3 September 1998 (for nine seconds), and one on 31 August 1999 (for seven seconds). This made

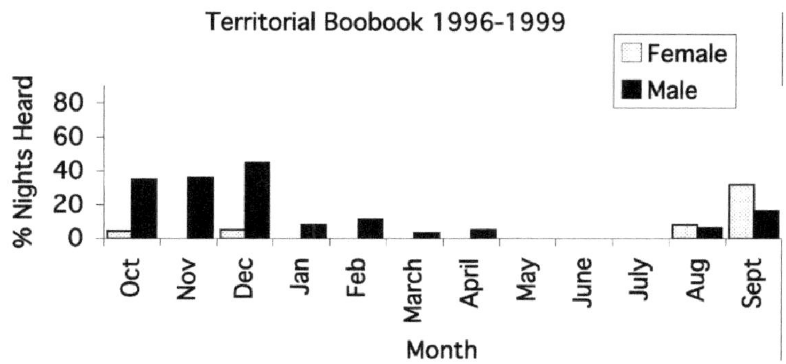

Figure 21.1 Per cent observation nights per month we heard Territorial Boobook callers identified as females (42/529 nights = 7.9%) (white) or males (103/529 nights = 19.5%) (black).

two bouts out of 67 for females, 3 per cent, and two bouts out of 188 for males, 1 per cent. The two bouts were brief and didn't fit Farabaugh's definition of duetting.

Precision of timing and sequential ordering

We tested Farabaugh's second and third criteria – *precision of timing*, whether the calls were so closely synchronised they sounded like one bird calling, and the *sequential ordering of elements*, whether the calls were coordinated in some fashion, *male-female-female-female*, or maybe *male-male-female-male-male-female*. But you can see in the percentages above that no pairs had overlapping calls for long enough to measure coordination, or sequential ordering of elements.

Commercial audiotapes had two Southern Boobooks calling at one time (overlapping bouts), usually induced by playing a Boobook recording inside their territory. We listened to determine if calls on the tape were synchronised (precision of timing), that is, if they sounded like one bird calling, and if the two birds called in an orderly fashion (sequence of ordering). But the series of calls fell out of phase so that deeper-voiced Boobooks on the tape started certain couplets, and higher-voiced Boobooks started other couplets. Though calling at the same time, the overlapping bouts sounded like two birds calling randomly, like two dogs in a yard barking at the mailman, not like the Everly Brothers. This, according to Farabaugh, was not duetting.

So, if the birds on the tapes were not calling synchronously, their calling was not orderly and mated pairs in the field almost never called at the same time, they were not duetting. What was the behaviour described by researchers? If it wasn't duetting, what was it?

Duels

We called the overlapping bouts between neighbours 'Duels' and defined Duelling as two birds from different territories facing each other, 1 to 50 m apart, and calling with Territorial Boobook calls, with bouts overlapping. It was interesting to watch how excited and aggressive these males became.

On 25 of 529 nights, (4.7 per cent), we observed males Duelling with neighbouring males. On one of 529 nights, (0.2 per cent), we saw females Duelling with each other, and on two of 529 nights, (0.4 per cent) we saw neighbouring males and females Duelling. At least 14 of the 28 duels (50 per cent) had three birds present, but only two of the birds actually called. Brent Stephenson in 1998 reported similar behaviour in Boobooks (Moreporks) on territory borders in New Zealand, and Hill and Lill (1998b) reported similar behaviour in Christmas Island Hawk-Owls.

It seemed that researchers had not seen duetting, they had seen either Duelling between neighbouring males or artificial behaviours caused by taped playback in the owls' territories. Because researchers had not sexed or colour-marked owls in their studies, they did not know the location of territory borders. When researchers heard two Powerful Owls calling, they often assumed that the pair was 'duetting' from the centre of their territory, but we wondered if Powerful Owls behaved in a similar fashion to their smaller Boobook cousins. The clusters on maps may not be mated pairs of Powerful Owls at territory centres, they may be opponents on borders.

Duelling was interesting. Why did males spend so much time in wars of attrition, hammering each other in vocal contests? Did Duelling affect each other's breeding? What were the costs and benefits to each male?

22
Weather and calling

Why do Boobooks call more often in some months and in some years than others? You can see this variation in Territorial Boobook calls in Figures 22.1, 22.2 and 22.3. The variation could simply reflect individual differences, except that we had watched the same owls over three years. This was calling by the same owls over a period of time, and the frequency of calling had changed.

Weather

Some researchers had claimed that weather caused this variation in calling. Boobooks were more likely to call on nights with little or no wind, when the moon was not visible, when the night was clear rather than overcast and on calm nights.

Another researcher contradicted this. Moon visibility and cloud cover, he said, did not affect calling rates, but rain and wind did depress calling, which declined from February through autumn into winter, then rose in late winter or spring. Calling was affected by weather, but calling was also seasonal.

We compared the frequency of Boobook calling to seven weather variables: (1) **moon visibility**, (2) **moon phase**, (3) **cloud cover**, (4) **rain** in mm, (5) **wind speed**, (6) **wind direction**, and (7) **temperature**. Then we looked at frequency in different seasons. We divided the year into four categories based on the breeding cycle of the owls – **winter** (1 March–14 August); **pre-breeding** (15 August–30 September); **breeding** (1 October–1 January); and **fledging** (1 January–28 February) (Olsen *et al.* 2002a).

We used Chi-square analysis and the Fisher's Exact Test to determine if calling differed according to the above variables. For example, did we hear calling more

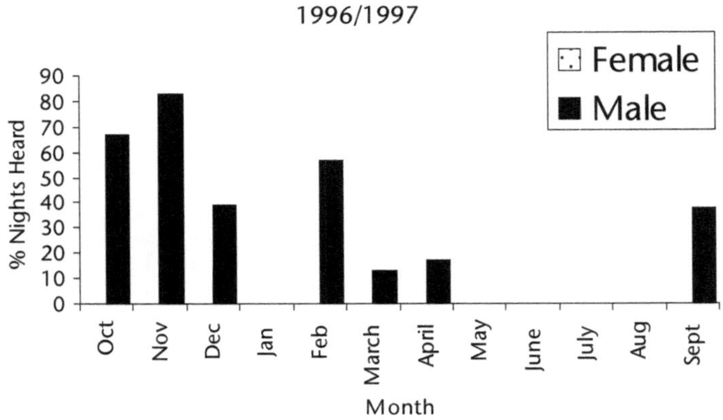

Figure 22.1 Per cent observation nights per month we heard Territorial Boobook callers identified as females (dots) or males (black) in 1996/1997. No females were heard.

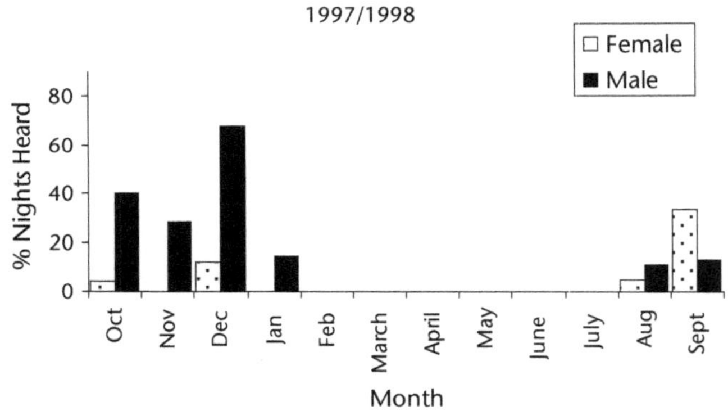

Figure 22.2 Per cent observation nights per month we heard Territorial Boobook callers identified as females (dots) or males (black) in 1997/1998.

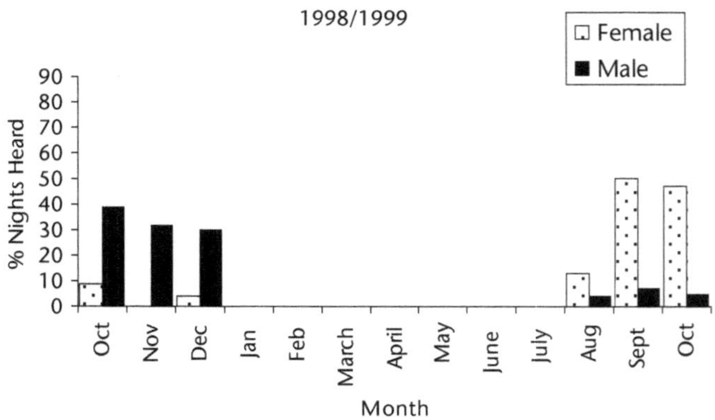

Figure 22.3 Per cent observation nights per month we heard Territorial Boobook callers identified as females (dots) or males (black) in 1998/1999.

often on nights when there was more cloud cover or less, when the moon was visible or hidden, when it rained or didn't rain?

Season was a powerful determiner of calling, but the other variables did not seem to matter. We found a significant difference between calling on warm nights versus cold nights, but weather and temperature were not independent. Boobooks called more on hot nights because they called more during the breeding season, and breeding started after nights started to warm; they called less often in winter because they had finished breeding and females were often gone.

Though heavy rain and wind might suppress Boobook calling as some studies suggested, it did not stop it. The calling was less *detectable* to researchers on rainy or windy nights because of swishing branches or drumming rain. This did not mean that owls did not call, it meant researchers could not hear them. We stood close to the owls we studied and were more likely to hear them during wind and rain. We did not play a tape, then stand back and wait for an owl to call in the distance, a protocol used in many studies, sometimes with a first listening phase when researchers tally unsolicited calls.

The claims that owls called more during a full moon, or during no moon, or when the moon was hidden seemed to contradict each other. Researchers had thrown so many variables into the calculating programs that 'artefacts' arose, differences that were not really differences.

Social context

The frequency of calling did vary from year to year, as you can see in Figures 22.1, 22.2 and 22.3. If weather did not cause this variation, what did?

Males often called during the spring and summer of 1996/1997. This included a number of 'Duels' between one male, Green, and his two neighbours. In 1993 Green seemed to start a territorial dispute with his two neighbours, Red and White. This lasted from 1993 to 1997, and he gradually wedged into their territories, took away their land. At the end of December 1997, one of the original males, Red, disappeared. Green seemed to have killed him or driven him off; maybe Red had died from all the stress. As you can see in the figures above, during the period following this, Territorial Boobook calling between males fell off because there were only two males left to call, White and Green. It seemed that three males had crowded into space big enough for only two, with enough food in the long term for only two nests, and this behaviour increased the amount of calling.

Three variables seemed to explain much of the variation in frequency of calling: (1) **season** – they called more in spring and early summer; (2) **individual differences** – some individuals called more than others; and (3) **social context** – if there was a border dispute, enemies to confront, a destabilised neighbourhood, a new nest, or a new border, males called more often. Challengers might kill each other if they had to.

After Male Red disappeared in 1997, there was an increase in female calling. Why did females increase calling after a dispute between males had finished? Females took part in these disputes, though marginally. We will discuss this further in Chapter 23.

23

A lover's triangle

Scientists assess the quality of habitat that raptors use. In theory, raptors settle good habitat before they settle poor habitat. When birds first settle an area, they settle in a sequence – good habitat first, moderate habitat next, poor habitat last. This is known as *sequential habitat occupancy*. For example, when birds first arrive back on their breeding grounds from wintering, or birds are introduced into an area for conservation purposes, the first wave of settlers takes up the best habitat and the last settlers take up poorest habitat. However, this assumes static habitat quality, that is, habitats do not change over time. It also assumes that individual pairs cannot change or improve the territory they settle on. Site fidelity, the tendency for owls to remain or return to one territory instead of moving from one territory to another, means that an owl can start breeding on good habitat and end up on poor habitat if this habitat changes or neighbours take away pieces of it. Below we look at a single example to illustrate what can happen when two pairs have settled on good habitat, and a new pair moves into adjoining poor habitat.

In 1993 males Red and White had been breeding successfully on adjacent territories when a third male, Green, moved to habitat adjacent to Red and White and bred. White and Red defended their western borders with territorial Boobook calling and attempted to push Green back, but Green stayed and continued to push east into Red and White's territories. For six years the three males battled over land.

Using bouts of calling on their common borders Green Duelled with his two neighbours. Apparently he sought the taller trees and better shrub cover that held food, call trees and nest sites on his neighbours' patches. Each year Green took a little more land from his neighbours. Then, at the end of December 1997 (see Figures 23.1 and 23.2), one of the losing males, Red, disappeared. Much of the vocal conflict between 1996 and 1997 happened on the border where Green pushed east into Red's territory. Red was apparently defeated or dead after nesting on the same land for at least six years. His enemy Green now had *Lebensraum*. This conflict was about limited resources and habitat quality.

Though Red Male disappeared from Aranda Bushland, his mate, Red Female, stayed. These three owls, Green Male, Green Female, and Red Female, were now on one expanded territory. Green shifted into the nest of the disappeared Red. He roosted where Red had roosted, in a high limb slanting over the nest, his face staring down from under the limb by day. At night he emerged and called.

Red Female stayed and floated around her invaded territory and one night in August, when the conquering Male Green sat and called from Red Female's old nest tree, she came in to him. He tried to mate with her but she flew off, maybe a little nervous about this new male, though she would have 'known' him for a number of years as the enemy neighbour.

The third owl in this triangle, Green Female, hadn't travelled over the hill with her conquering mate. She stayed in her wintering tree west of the ridge and hunted in suburbs until, one night, she moved east over the hill, closer to Green and closer

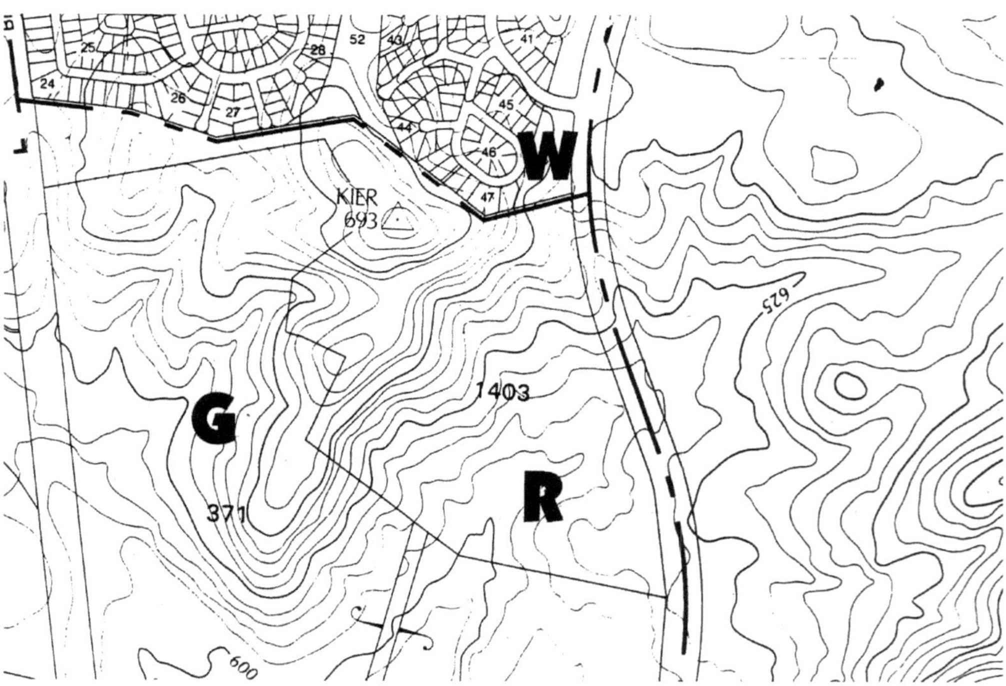

Figure 23.1 Nest sites of males Green, Red, and White in 1993. Green Duelled on his border against White and Red.

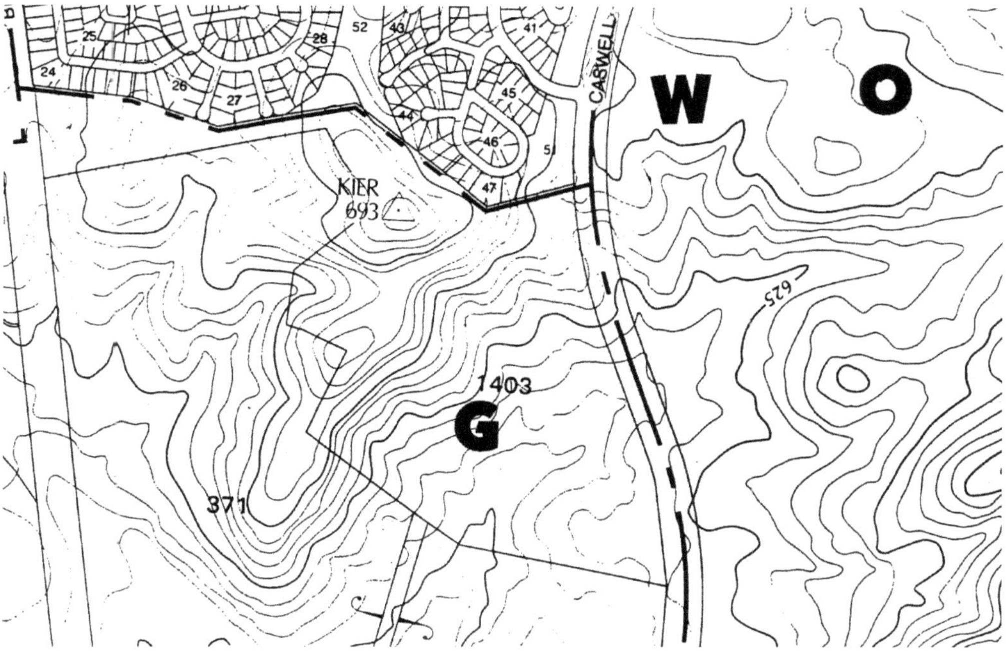

Figure 23.2 Nest sites of males Green and White in 1998, after Red disappeared and Male Orange moved its nest south.

to Red Female (see Figures 23.1 and 23.2). We now had two females on one territory, or perhaps Green had two females on his new double territory.

The forest on Red's side had two big Brittle Gums – the nest tree and a second high tree that owls used for calling. After Green Female arrived, she sat and Boobook called in a long bout from the second big tree, but Green sat next to Red Female and watched his old mate. The two females challenged one another in a vocal duel, then, surprisingly, Green flew at his former mate perhaps to mount her and show his preference, or maybe drive her away. She flew over the rushing creek into forest.

The following night, September 1, 1998, the Boobook threesome had decisions to make if they would start a breeding cycle in time to produce chicks that season. Green now lived permanently on the shady east side of the ridge.

Green emerged but he didn't call. Another Boobook called to the west, high up on the hill about 100 m away, downhill from the radio-tagged Green Female. Green Male flew lazily away and drifted over the rushing creek and croaking frogs.

Apparently the two females, Red and Green, were alone on the hillside.

The Boobook calling from the hillside stopped, and the woods went quiet. Green Female's radio signal drifted downhill at an angle towards the Red Female, to a shadowy ravine.

There was no more calling.

Green Male stayed put. White moonlight leaked through the forest canopy. The radio signal bounced around in the trees, especially in steep gullies, and made it hard to locate the radio-tagged Green Female.

Sue pointed to a form on a branch some 10 m above us. 'Is that her?' she whispered.

Lacy black shadows spread over the gnarled Brittle Gum where Green Male and Female had roosted the previous winter on the border with their enemies, the Red pair. A long thin tail hung down like a Sugar Glider's tail but it seemed the wrong shape. Green Female looked into the torchlight and closed one eye. She often did this, probably because she nursed an injury to that eye. She was the biggest and most powerful Boobook in the study; she lacked the agility to navigate through dense forest at night. We had seen her hit branches.

Green Female sat on a large body. She flapped heavily, carried it 10 m to another branch and alighted. We took care not to break any sticks and frighten her, and shone the light again. She sat on a full-grown owl. She had stripped the secondaries and most of the primaries from one wing so the hanging wing looked like a tail.

She flew again and because of the weight of her quarry, she landed on a platform of three dying trees leaning into each other. The head was gone, a bloody stump poked out of the feathers, like a thin, red chicken neck.

It seemed that Green Female had called from the hillside with Territorial Boobook calls as a challenge to other females, and Red Female had drifted down and found her. Green Female had killed her.

The following morning we searched for the severed head but she must have stashed the remains in a tree hollow, or maybe dropped them to the local fox.

That night the Green pair called softly to each other.

The 1st of September 1998 marked the end of a six-year battle over land in the forest. The Duelling stopped, then it started again in 2002.

In this case, owls Red and White had settled good habitat first. Then Green arrived. He took away pieces of territory from Red, then from White, to improve his own territory. Green took more from Red each year over five years, then finally killed or expelled the Red male, creating an expanded territory. The density of breeding pairs on this patch went from two pairs in 1992 to three pairs in 1993, then back to two pairs in 1998, with no obvious change in prey availability or nest sites. Owls had changed the pair densities in this forest.

24
Pre-breeding behaviour

As mentioned above, the reproductive period of owls is prolonged compared with that of many birds. Even the Southern Boobook, our smallest owl, takes at least five months to complete a breeding effort. Most owls establish nesting territories large enough for breeding and for both the male and female to hunt in. Site fidelity is high in *Ninox* owls. One or both members of the pair often remain year-round on the territory, or return to the same home range each year. Owls such as Barking Owls and Powerful Owls are said to form monogamous, lifelong bonds but this seems unlikely and needs confirmation. Long-term study is necessary. Pairs can remain bonded for many years, or until one dies, but studies of territories over many years, even decades, probably will show that some individuals switch territories and mates. More likely, females change territories more often than males, and home range sizes change over time (Olsen and Trost 2007).

During winter, the male Southern Boobooks we studied remained on their breeding territories, but females often travelled 7 to 10 km away and spent the winter on a separate home range, or sometimes wintered in the far corner of their old breeding territory. We will discuss 'female desertion' in more detail in Chapter 30. During winter, females often returned to the breeding territory and to last year's male for a night or two, and roosted with him. Returning females mutual preened with their mates and used low croaking calls in greeting, then returned to their winter home range. The female in Table 24.1 returned on 14 September and stayed for the rest of the breeding season.

Boobooks use territorial calling through much of the year, but nights of calling are less frequent during winter, then they abruptly increase during spring

Table 24.1 Dates during 2000 when one radio-tagged female Southern Boobook was seen in her breeding and winter home ranges, 8 km apart in Canberra (Olsen and Taylor 2001)

Dates	Home range
7 June to 31 July	Non-breeding
4 to 8 August	Breeding
11 August	Non-breeding
14 September	Breeding

(September–December), sometimes close to the nest, but sometimes away from the nest along territory borders, apparently aimed at neighbours. After females returned to a male's territory for the breeding season, often last year's male but sometimes to a different territory and different male, both sexes increased their Territorial calling (Figure 24.1).

Spring 'floaters', unmated owls with no breeding territory, pass through looking for vacancies and breeding opportunities and they sometimes call. Males especially can drive away these intruding males or females and resident females can do the same.

Neighbouring owls might respond to loud territorial calls, and males and/or females might travel to borders and challenge them. Females give territorial calls less often (Figure 24.1), in Boobooks a third as often as males did in our studies, but they did give them, particularly in the pre-laying phase. Maybe females are stamping their ownership onto a nest site, signalling to other females to stay away. As mentioned above, Boobooks do not duet as is sometimes reported in the literature, but their bouts of calling occasionally overlap those of males.

On rousing at dusk during pre-breeding, owls mantle and stretch. As the sky darkens, some maintain contact with bouts of 'Contact' calls, muffled versions of the territorial *Boobook* of Southern Boobooks or the *Who-who* call of Powerful Owls.

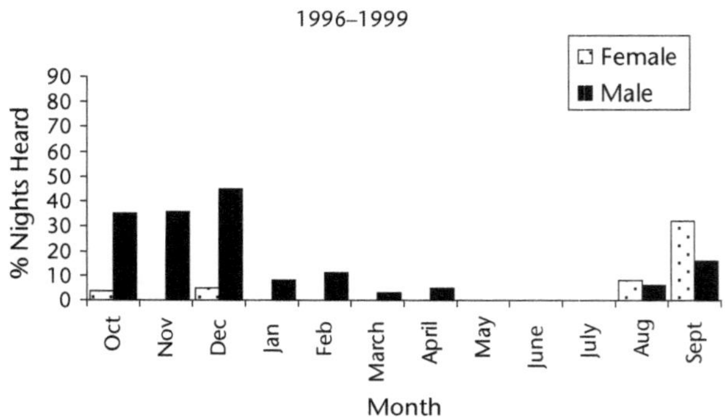

Figure 24.1 Per cent of observation nights per month that we heard Territorial Boobook calls from 1996–1999 that we identified as females (dots) 42/529 nights (7.9%) and males (black) 103/529 nights (19.5%).

Females also start Bray calling, to induce males to provision them. Males may give bouts of Territorial Boobook calling. These calling patterns help drive the breeding system of owls – males competing for females and females choosing quality males. Most Bray calling, or its equivalent, 'Bleating' in Powerful and Rufous Owls, occurs before egg-laying and during incubation and brooding, when males spend their nights hunting for the female and brood. You can see this in the annual timing of these two vocalisations (Chapter 19). Male Boobooks call in spring to stamp their ownership on a territory, advertise and bring in females to sample food and nest sites there. A female Bray calls to induce a male to forage for her and 'prove' to her that he is a good provider. If he fails this test, or the female 'dislikes' something about the territory or the male, she moves on and tries again, or does not breed that year.

In the courtship period, some of this behaviour, including calling, allopreening and roosting together, occurs on borders or in different parts of the territory, but much of it occurs around the nest. Courtship proceeds in *Ninox* with long bouts of croaking calls (often at the nest hole), and allopreening. The strong shows of 'affection', with pairs face-preening their mate (see Figure 24.2) and sitting with bodies touching, appear strange to anyone who has studied certain diurnal raptors such as falcons or goshawks. In many diurnal raptors the sexes rarely touch each

Figure 24.2 Owls show affection to each other during the pre-breeding period.

Figure 24.3 Southern Boobooks mating.

other and males often seem frightened of females. This underscores the phylogeny of different raptor groups – raptors may all kill vertebrate prey, but this is convergent evolution – different basal groups of birds evolving to capture and partition vertebrate prey. Owls seem to be related to woodpeckers, but falcons are related to passerines.

The male provides food for the female, and she continues to beg with Braying or 'Bleating' calls. These and other calls accompany mating, on a branch near the nest (Figure 24.3), and culminate in a Mating Squeal – a call that sounds remarkably like a rabbit being caught by a predator.

Less is known about pre-breeding behaviour in Australian *Tyto* owls. Though *Ninox* owls are seasonal, some *Tyto* owls are said to breed at any time of the year when food is abundant, although there is an underlying broad seasonal trend in large forest *Tyto*. David Hollands has described pre-breeding in some *Tyto* and they appear to perform noisy display flights, often near the nest, behaviour unlike that of *Ninox* species. Courting *Tyto* males may circle near the nest and give screech and chattering calls before mating with females. There is much work to be done on the behaviour of the world's poorly known *Tyto* owls, many of which live in the Australasian region. We have made a better start with Australia's *Ninox*.

Plate 1: Tytonidae owls, such as this Pacific Barn Owl, have a distinctive, heart-shaped facial disc.

Plate 2: This starling-sized Eurasian Pygmy Owl in Finland is a diurnal species that preys on birds and invertebrates.

Plate 3: Southern Boobooks in Tasmania are coloured differently from Boobooks on the mainland and some taxonomists consider them a separate species. Because of a shortage of hollow-bearing trees, Boobooks on King Island often roost and nest in human-made structures.

Plate 4: Powerful Owls have massive feet and beak for capturing, killing and dismembering large prey.

Plate 5: Fledged Southern Boobooks roosting in Native Cherry.

Plate 6: Southern Boobooks in nest hollow.

Plate 7: Day-roosting juvenile Southern Boobook playing with its sister's toe.

Plate 8: Flip-trap used to capture owls.

Plate 9: Trapped adult Southern Boobook in Canberra.

Plate 10: Adult Little Owl.

Plate 11: Trapped and patagial-tagged Snowy Owl on the Skagit Flats.

Plate 12: Colour-banded Southern Boobook arrives at nest hollow with a Christmas Beetle.

Plate 13: Jemima Parry-Jones with Eurasian Eagle Owl.

Plate 14: Species such as the Eurasian Eagle Owl commonly nest on the ground.

Plate 15: Though fledgling Southern Boobooks did not hunt vertebrates, their parents fed them birds, bats and other vertebrates.

Plate 16: Juvenile Boobooks in Tasmania have yellower eyes and whiter feet than do Boobooks on the mainland.

Plate 17: On cold nights when insects were not moving, adult Southern Boobooks brought in more birds to their fledged young.

Plate 18: Masked Owl.

Plate 19: Fledgling Southern Boobook swallowing leg of a Common Myna.

Plate 20: On hot nights female Southern Boobooks may go off to bathe while males hunt for prey.

Plate 21: Juvenile Southern Boobook consuming an insect brought by its mother.

Plate 22: Tawny Owl in Spain (Iñigo Zuberogoitia).

Plate 23: Little Sumba Hawk-Owl, a new species.

25

Nests and breeding habitat

As mentioned above, owls build no nest. Owls in Europe often use nest boxes, but Australian owls seldom do. Australia still has many large trees and most Australian owls, except Grass Owls, choose a nest hollow in the trunk or limb of a tree. They seldom use nest boxes except in places such as Melbourne or King Island, Tasmania, where Powerful Owls and Southern Boobooks respectively do use them. In Europe, where forests were long ago felled, entire populations of owls breed in nest boxes nailed or wired to trees, giving easy access for study. Researchers do not search for these nests because the researchers put the nests there in the first place. Not in Australia.

North American owls such as Great Grey Owls or Great Horned Owls often take over the nest of a hawk or raven, a nest clearly visible from above or below. In places like the dry eastern border of Washington State, you can find Great Horned Owls in the abandoned stick nests of Red-tailed Hawks in Ponderosa Pines, easily seen from the ground. Brooding or incubating females are conspicuous to observers and the adults stand in bright sunlight next to their open nest and young. Australian owls incubate and brood in deep hollows or caves, often in dark forest; they almost never use stick nests. Compared with owls in Europe and North America, Australian owls are invisible.

Powerful Owls

Tall, shaded forest is typical Powerful Owl country. They are said to hunt in open eucalypt, or 'sclerophyll', forests in particular. Like Powerful Owls, sclerophyll plants are adapted to Australia. They have leathery leaves high in oil, with fewer stomata than other plants, and their leaves mostly hang vertically and twist on stems so they avoid too much exposure to the sun. This reduces water loss. These sclerophyll forests support prey species that Powerful Owls commonly feed on, arboreal mammals that eat leathery leaves, fruits and lick sap from wounds in trees. Some of these mammals hunt through eucalypt bark and canopy leaves for insects. Powerful Owls prefer to hunt them in the canopy of open forest, but prefer to roost and breed in denser forest, in dark sheltering gullies.

Rod Kavanagh, Stephen Debus, Ed McNabb, Raylene Cooke, Rainer Rehwinkel and others have assessed this habitat. The tree species[1] that make up the open and closed forests that Powerful Owls live in have common names that are musical to the ear, and scientific names that roll off your tongue. At Kioloa on the south coast of New South Wales, Powerful Owls live in tall forest dominated by Spotted Gum *Eucalyptus maculata* and Blackbutt *Eucalyptus pilularis*. In Victoria, they use box and ironbark forest, Mountain Grey Gum *Eucalyptus cypellocarpa*, Mountain Ash *Eucalyptus regnans*, Candlebark *Eucalyptus rubida*, Messmate *Eucalyptus obliqua* and River Red Gum *Eucalyptus camaldulensis*.

In the ACT we found them in forests dominated by Manna Gum *Eucalyptus viminalis*, sometimes called Ribbon Gum because long strips of shredded bark hang like Christmas trimming in the elbows where limbs emerge from trunks, and out along the forks. Powerful Owls also like Brown Barrel *Eucalyptus fastigata*, Narrow-leaved Peppermint *Eucalyptus radiata* and Mountain Gum *Eucalyptus dalrympleana*. For daytime roosts they sometimes use Native Cherry *Exocarpos cupressiformis* – dark green evergreens that resemble squat, sculptured Christmas trees, like the onion-shaped domes in the Kremlin – or Blackwoods *Acacia melanoxylon*, Lilly Pilly *Acmena smithii*, pittosporum species or introduced conifers.

Powerful Owls nest in large vertical hollows, typically broken-off branches in trees or natural spouts that lean out horizontally or angle upwards from the trunk. Eggs and nestlings rest on wood dust and chips in the bottom of the hollow. The nest tree is often near a permanent creek, on a sheltered slope instead of a high exposed ridge.

Ed McNabb (1996) measured 21 nests near Melbourne. Almost all nest hollows were in living trees within or below the forest canopy and thus sheltered by it. McNabb estimated the age of these trees as 150 to >500 years old. Mean height of entrance above the ground was 22 m (8 to 40 m). Nest hollows in the smaller (younger) trees were used a mean of one time, taller (older) trees 1.8 times. Twenty

1 From Davey 1993; Kavanagh and Peake 1993; Kavanagh and Bamkin 1995; Higgins 1999; Traill 1993; Olsen and Rehwinkle 1995.

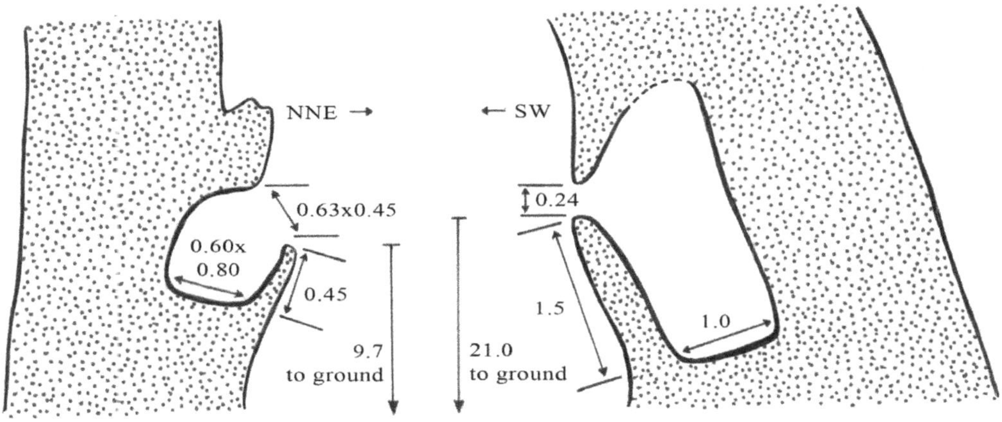

Figure 25.1 Measurements (metres) of two Powerful Owl nests.

of 31 sites were beside or within 100 m of permanent water. Nest hollows were oriented to protect them from inclement weather.

In New South Wales Chafer (1992) and Debus and Chafer (1994) measured or collated data on eight nests, and found a mean height of 23 m (9 to 37 m); depth of nest hollow (internally from entrance to the chamber floor) was 45 to 180 cm; and diameter of nest hollows at the chamber floor (diameter of the egg chamber) was 46 to 100 cm (Higgins 1999).

One Powerful Owl nest on the New South Wales coast fitted these descriptions. When John Kershaw, Ian Rowles and I arrived on 31 August, the adults flushed, leaving their two fledged young behind on a limb. The young Powerful Owls looked like giant fluffy Southern Boobooks, but the size of hens. They had the same fluffy cream plumage, brown wings and raccoon masks as Boobooks, but a more serious expression, a more formidable look.

As I climbed to the nest, the larger adult male returned and perched close to the fledged young, and glared as if threatening me. The nest entrance measured only 9.7 m from the ground, the cavity opening was 63 by 45 cm across, and facing north-north-east; it was half a metre deep, and 80 cm across the bottom, plenty of space for the family as it grew, on an east-facing hillside in Hawkesbury Sandstone country. The nest dimensions were a little different from a nest in Namadgi National Park (see Figure 25.1).

Ed McNabb and Jim Greenwood near Melbourne have tested various designs in areas where Powerful Owls have enough prey but lack suitable natural hollows. Dimensions of boxes used by the owls at Blackburn were: 700 mm high at the front, 800 mm high at the back, 500 mm by 500 mm square at the bottom with an entrance hole 200 mm wide and a sloping roof. The base was domed for drainage and was removable. Pieces of wood covered the floor as substrate.

Southern Boobooks

As mentioned above, Southern Boobooks nest where there is woodland, in all the habitat mentioned above for Powerful Owls and a lot of habitat that Powerful Owls cannot use, for example, all of Tasmania and creek-lined desert regions across the inland. In Canberra breeding pairs nest in open forest and tall woodland, dominated by Scribbly Gum *Eucalyptus rossii*, Brittle Gum *E. mannifera*, Red Stringybark *E. macrorhyncha*, and Blakely's Red Gum *E. blakelyi* woodland with Red Box *E. polyanthemos* and Yellow Box *E. melliodora*, with an understorey of tussock grasses (*Poa* spp.) and the shrub *Cassinia longifolia* dominating in more open areas. A regime of prescription burning has created a mosaic on the understorey where Boobooks hunt along forest edges, roadsides, bushland corridors and in backyards. They commonly day-roost in Native Cherry, bushy conifers and deciduous trees.

Boobooks compete with honeybees, arboreal marsupials, cockatoos, kookaburras and a number of other birds and mammals for nest hollows. It is noteworthy that we have better data on nest-site characteristics and prey of Powerful Owls than we do on Southern Boobooks. Higgins (1999) summarised measurements mainly from the Nest Record Scheme as: height of entrance above the ground – 7.2 m (3 to 30 m, n=25), height of the nest tree – 12.3 m (8 to 30 m, n=12); depth of nest hollow – 30 to 120 cm (no n given); diameter of nest hollows – 20 to 30 cm (no n given), but we suspect that many nest entrances are smaller than this, to exclude large cockatoos and possums. One old stick nest was used 11 m above the ground (Sedgewick and Morrison 1949), but this is an old record and needs to be treated with some caution as some early observers confused Boobooks with Tawny Frogmouths that do build a stick nest, or it could be another species altogether.

Large forests are peppered with likely nest hollows – in tree trunks, empty dead limbs, broken-off spouts – holes just big enough to allow the owls in but keep a cat or Common Brushtail Possum out, about the size that would allow passage of a champagne bottle. Measurements of two Boobook nests we climbed to on Black Mountain in Stringybark/Scribbly Gum woodland are shown in Figure 25.2.

Though these two nests faced inclement weather to the north-west, they were shielded from storms by surrounding forest. The two nests had smaller nest entrances than those suggested by Higgins (1999), but some Boobook nests in the study had larger nest openings and the owls had to fight off Brushtail Possums that entered.

Southern Boobooks on King Island, Tasmania, regularly use nest boxes built by Nigel and Mavis Burgess from 30 mm thick pine driftwood (Figure 25.3). The square hole is 120 mm × 70 mm but can be altered. These boxes are 500 mm high

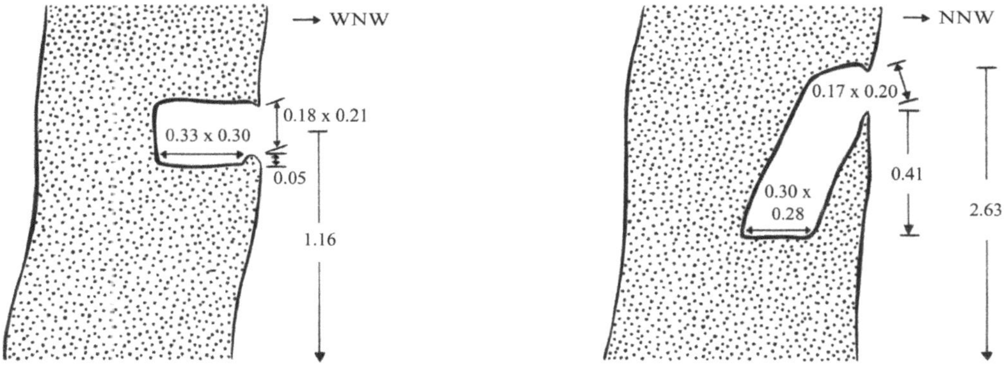

Figure 25.2 Measurements (metres) of two Southern Boobook nests.

and 250 mm square at the bottom with a sloping 300 mm roof and a 250 mm × 150 mm ledge at the front for owls to land on, and a 500 mm long stick nailed

Figure 25.3 Dimensions of nest boxes (mm) built by Nigel and Mavis Burgess for Southern Boobooks on King Islanc, Tasmania.

under the ledge for owls to perch on. All of this hangs from a 15 cm nail hammered into a large eucalypt trunk.

Restoring habitats, for example constructing wooden nest boxes to increase the numbers of breeding pairs, normally means that owls are retained in dramatically altered environments from the environments they evolved in. This is the case on King Island where vegetation was cleared after World War II and settlers were offered land there. Now the island is dominated by grazing land and *Melaleuca* trees that rarely form suitable nest hollows for owls. Nest boxes can play an important role in species and population conservation as long as the chosen sites are protected from inclement weather, direct sunlight, predators and human interference.

Nest boxes are no substitute for maintaining natural sites in natural habitat – natural tree hollows and woodland. Nest boxes have become the norm in places like Europe, used in place of restoring natural habitat. In Australia, they should only be used as a last resort.

Higgins (1999), Hollands (2008), Debus (2009) and the Field Guide below (Appendix A) contain data on nesting habitat and nest-site characteristics for other Australian owls.

26
Timing of breeding

To understand the breeding strategies of owls we need to understand why they breed when they do, the timing of breeding. Compared to Europe and North America, fewer owl nests have been systematically watched in Australia. But even when David Fleay followed owls in Victoria in the 1940s there were clues about why Southern Boobooks laid eggs and fed nestlings in spring, and why Powerful Owls laid eggs in autumn and fed nestlings during winter – Boobooks feed their young many invertebrates, Powerful Owls feed their young many arboreal mammals.

Timing of breeding in Southern Boobooks

Southern Boobooks around Canberra start breeding in late August and fledge young mostly in December. This timing is less of a conundrum than is the timing of Powerful Owls because Boobook breeding is in line with most other Australian raptors and in line with many Australian birds. Many raptors in south-east Australia breed in late winter/early spring, and fledge their young in late spring/early summer, commensurate with a flush of available prey. Ornithologists find it useful to separate 'ultimate' factors, the *reason* why owls breed when they do, from 'proximate' factors, the actual *triggers* that cause owls to start breeding. Ultimate factors, the ultimate reasons for having eggs, nestlings or fledglings at a certain time of the year, can include food supply in a particular season. It is a convention in the birds of prey literature that the ultimate factor in the timing of raptor breeding is so peak food supply occurs during the nestling period (Newton 1979). However, Olsen and Georges (1993) argued that ultimate timing might relate to the period when young raptors have left the nest and are learning to hunt. Most young raptors, when they fledge, cannot successfully capture enough prey to sustain themselves, so parents continue to feed them. Before their parents stop feeding them, fledglings need to practise capturing birds, mammals or insects that are their main prey, otherwise they will starve. This post-fledging period provides practice in hunting.

Newton (1986) argued for Sparrowhawks that the peak in vulnerable just-fledged passerine birds was the ultimate reason for the timing of breeding; just-fledged passerines are easier to capture than are adult passerines, so adult Sparrowhawks capture these easy prey to feed their dependent young. However, an examination of Newton's (1986, Figures 2 and 39) data shows that the peak of fledged passerines occurs between the time when young Sparrowhawks begin to fledge and begin to disperse, that is, the time when they are learning to hunt but are still being fed by adults. This is not solely a time when adults are hunting for dependent young, it is a time when young are learning to hunt.

The ultimate reason why Southern Boobooks breed when they do may be the flush of flying insects and ground invertebrates, such as spiders and cockroaches, that occur each summer (see Chapter 10). Even clumsy young owls can take them, and young Boobooks hunt them almost exclusively between the time they fledge and disperse. We never saw a fledgling take vertebrate prey, but adults, experienced at taking vertebrate prey, did catch parrots and other birds to feed fledged young. Insects contain fat, they are nutritious. Our colleague Tony Rose told us that soldiers in World War II in tropical regions were advised to eat insects because they were such a good source of nutrition and fat. Heat is a key to this timing of breeding. Invertebrates move and fly around more in warm months than in cold months, when they are more visible and vulnerable to capture by owls. Invertebrates are also easier to capture than are vertebrates, such as birds and bats.

Proximate factors

Proximate factors are the actual triggers that cause owls to start breeding. Factors such as day length may be the initial stimulus, along with enough food, to kick off breeding in late winter. In south-east Australia male and female Boobooks start territorial calling, and females start 'Bray' calling in late winter/early spring. Prey abundance for Southern Boobooks goes through a bottleneck. It can be lowest in winter when spiders, insects and birds are least available, and reach a peak in late summer when Coleoptera (beetles), Lepidoptera (moths), spiders and the fledged young of parrots and small passerines are most available. At the end of summer food supply declines again and becomes scarce, becoming a bottleneck over winter. This is a potential problem. Owls need enough food to bring them into breeding condition. Close to the winter low in food supply, Boobooks start pre-breeding behaviour so they can fledge their young by the summer peak in food supply.

With many birds of prey, changes in day length stimulate gonadal growth and increase in female appetite. She responds by catching more prey and accepting extra food brought by the male. She lays down fat reserves and will experience growth in both the ovary and oviduct, and begin yolk deposition. Males will experience growth of the testes, but only female owls lose feathers from their lower abdomen, an area of vascularised skin through which heat is transmitted to eggs and nestlings during incubation and brooding. Male owls do not incubate, and are not known to brood.

Though season and increasing day length were likely proximate causes for breeding, Boobooks also took advantage of certain waves of food that passed through their territory in late winter/early spring, for example Bogong Moths *Agrotis infusa*, a good source of fat for the owls. They also took House Mice that often bred in winter, as well as small birds. That is, there is flexibility in proximate timing of breeding, and this may link to weather, and the availability of invertebrate and vertebrate prey. As mentioned above, fledged young only hunted invertebrates (see Figure 26.1), such as flying insects, and these were most abundant in December–January. These abundant summer invertebrates, as well as the coming bottleneck of low food during late autumn and winter, were the likely ultimate reasons for the timing of breeding. Increasing day length and short flushes of prey such as Bogong Moths, mice and small birds are the likely proximate triggers.

Figure 26.1 Fledged Southern Boobook learning to take insects from eucalypt leaves.

Powerful Owls

Powerful Owls take much heavier prey than do Boobooks, and proportionally fewer invertebrates. They are strong enough to kill the family cat, which they sometimes do. But, as Fleay (1968) said, their mainstays, their four favourite prey species are Common Ringtail Possums, Common Brushtail Possums, Greater Gliders and Sugar Gliders. Though Powerful Owls eat a variety of prey including birds and bats, I limit the discussion here to these arboreal mammals because Powerful Owls do successfully breed without bird and bat prey, but there is no evidence that they breed successfully without at least two of these four arboreal mammals as prey. Furthermore, many owls around the world take bats and birds, including parrots, but do not initiate breeding in autumn. These four arboreal mammals better discriminate Powerful Owls as predators from other raptors, and Powerful Owls are most common and most successful where these mammals occur.

Powerful Owls breed in autumn, with many clutches in south-east Australia laid in late May or during June. They are feeding nestlings around July/August in high winter, and fledge young around August/September, so juveniles are learning to hunt in spring and summer, before they disperse in late summer or autumn. Are adults feeding nestlings, or are juveniles learning to hunt during peaks in the Powerful Owl's four main prey?

1 – Common Ringtail Possum. These strictly nocturnal possums are leaf-eaters that also feed on eucalypt flowers and fruits. They are infamous rosebud thieves in Victorian suburbs and they breed from autumn to spring, often giving birth at the beginning of this season. The females have four teats and usually carry two offspring, and these adult females with pouched young would be available for Powerful Owls in mid-winter to feed their nestlings. Young possums leave the pouch when they are about four months old, around August, when they are probably at risk from prowling owls. The young cling to the mother's back while she forages, and they are weaned at about six months old, around October. So this is the time that juvenile Ringtail Possums just learning to forage for themselves become available to fledgling Powerful Owls that are learning to hunt.

2 – Brushtail Possum. This prey species occurs across the Powerful Owl's breeding range; they are the heaviest of the four main prey species, at 1500 to 3500 g for females and 4200 to 4500 g for males. Tilley (1982) noted that Powerful Owls took juvenile Brushtail Possums, and Tony Rose (1993) showed that Powerful Owls do take some adults. Adult males are probably not taken.

Brushtail Possums occur wherever there are trees and can be a common species in urban areas. Although the liver of these possums can detoxify the poisons in eucalypt leaves to a point, they cannot cope with an exclusive diet of eucalypt leaves so they supplement this diet with fruits, buds and some pasture plants, as well as meat and eggs. Births are recorded in all months of the year, but they have a major autumn and a minor spring breeding season. Young may spend four to five months in the pouch, and a further one or two months riding on the females' back. Lavazanian *et al.* (1994) found that Brushtail Possums were more likely to be taken by Powerful Owls in spring than in other seasons, and that these were most likely juveniles.

3 – Greater Glider. This is the largest of the gliding possums. They live almost exclusively on eucalypt leaves, as do Koalas whose young are snatched off their mothers' backs by Powerful Owls. Powerful Owls tend to take Greater Gliders at high altitudes, and less often or not at all at lower altitudes.

Greater Gliders breed between March and June and females have two teats, but only one young is born. The young emerges from the mother's pouch when three or four months old, around June to October, when it would be vulnerable to attack from some breeding Powerful Owls feeding nestlings and some fledged owls learning to hunt. The young gliders can stay in the nest or be carried on the mother's back for another three months before earning their independence at about nine months old, around December to March, when juvenile Powerful Owls are gaining independence.

4 – Sugar Glider. This small glider is common in forest and in some woodlands. Sugar Gliders achieve independence around March to May, making them available to hunting Powerful Owls in autumn, when females need fat to begin a breeding cycle. Where the forest has plenty of tree hollows and there is plenty of food, such as insects and the sticky gum leaking from scars in *Acacia* trees, or certain eucalypts, at least 10 Sugar Gliders can live per hectare of woodland, though more often densities are lower. Groups of up to seven adults and their young sleep at times in a common nest. They recognise each other by odour and they scent mark their territories against other groups. Females normally produce two young around August that remain in the pouch for about 70 days, to around October, and then mothers leave them in a group nest for another 30 days to about November. These young leave that nest at about 100 to 110 days old to forage, usually accompanied by their mother. The adults drive out most of these young when they are 7 to 10 months old, around March to May. The young may set out across open country to reach other forest so in these first months of independence you would expect mortality to be high, particularly from owls. Females can breed twice per year, winter or spring, so young can also become independent in late spring–summer.

Male Sugar Gliders may stay on a territory after they are grown, baby-sitting their younger brothers and sisters, keeping them warm and guarding them from predators. The young male may scent mark the territory for his parents, and cut into the bark to allow sap to flow, and he may one day inherit that patch of forest from his parents, if he lives that long. (The closely related Squirrel Glider *Petaurus norfolcensis* is important prey in some regions.)

We have to make an assumption here, that young of these four main prey species just out of the pouch, or just independent from their parents, are more vulnerable to attack from Powerful Owls. From December to March independent young Powerful Owls learning to hunt can take young Brushtail and Ringtail Possums riding on their mother's back or newly independent, and also take Greater Gliders just out of the pouch. Lavazanian *et al.* (1994) found that Powerful Owls took Ringtails through the year but took proportionally more juvenile Ringtails in summer just before juvenile Powerful Owls dispersed and they took fewest in spring. Sugar Glider young out of the nest around November, and displaced from their natal territory, could move to unfamiliar ground and be more vulnerable to owl attack around March to May. Or perhaps adult female possums and gliders are slow or clumsy when they have young in the pouch and these adults are more prone to attack during autumn. Lavazanian *et al.* (1994) found Sugar Gliders and Ringtails were equally taken in all four seasons but Brushtail Possums were taken mainly in spring.

How does this tally with the timing of Powerful Owl breeding? They lay eggs in autumn from May to early June, and are said to incubate these for 35 to 40 days. They feed their nestlings during high winter, July to September, the nestlings take seven to eight weeks to leave the nest around August to October, still covered in fluffy down as 'branchers'. They roam about, settling on limbs near the nest and flapping their wings before they can fly well.

Slowly, they gain the power of flight, all the while calling in a monotonous sort of trilling that lets parents know they are hungry. This bleating keeps the family together as they move through the forest. The fledglings are said to remain dependent like this, following parents through the forest at night, begging, roosting with one parent or the other for six to seven months, August to January. They may still receive food from their parents, even in the latter stages, and they may start to hunt for themselves and roost away from their parents at 12 to 14 weeks old.

This timing makes sense when you consider the times of the above four marsupials leaving the mother's pouch, or out of the nest and reaching independence, the time when these marsupials have left the supervision of experienced adults and are learning to forage for themselves on a strange territory. Young owls might chase, miss, chase them again but gradually learn how to capture them. This is a dangerous time for the young owls and for their prey, because the owls may be more prone to accidents from collision or starvation, and the possums and gliders more likely to be caught.

Cooke *et al.* (2006) showed that overall prey availability of these four main prey species (not just juvenile availability), determined through spotlighting in Powerful Owl breeding areas, was highest in spring, when fledged Powerful Owls were learning to hunt, and lowest in autumn when Powerful Owls were starting the breeding cycle. Prey availability did not peak when Powerful Owls had nestlings.

What are the proximate and ultimate triggers to breeding? Young Sugar Gliders, achieving independence around March to May, are available to adult Powerful Owls *starting* a breeding cycle, so decreasing day length (not increasing day length as in Boobooks) and abundant Sugar Gliders as well as other prey are likely proximate triggers. Availability of birds and insects may also be important. Sugar Gliders breeding in winter or spring are available to young Powerful Owls learning to hunt in spring–summer. If the breeding cycle is 10 or 11 months long, Powerful Owls no sooner rid themselves of dependent young before they start the cycle over again. Territory or home range is a spatial concept, but also a temporal concept. Powerful Owls need territories large enough to contain enough prey for them to rotate from species to species through the year as different prey becomes available, maybe in different parts of the territory.

Some raptors, it seems, time their breeding so they have nestlings, not fledglings, when food peaks (Newton 1979; Holthuijzen 1990). It is reasonable to predict that different strategies apply to different species, and important to point out that Powerful Owls breed when days are shortest and nights are longest, thus increasing foraging time for a nocturnal hunter. However, a key to understanding the timing of breeding in many raptors may be the length of the fledgling stage of breeding, when juveniles are learning to hunt. The longer this fledgling period, the earlier the raptor will breed. Raptors breed early enough to complete a breeding cycle.

27

Pair behaviour – incubation

As mentioned in Chapter 24 the breeding period of owls is prolonged compared to other birds. Even the Southern Boobook, our smallest owl, takes at least five months to complete a breeding cycle, about the same amount of time as a Peregrine Falcon. Southern Boobooks have incubation periods around 30 days, nestling periods around 42 days, and post-fledging periods around 30 to 60 days, similar to a Peregrine.

Most owl eggs are white, not coloured as falcon, eagle or hawk eggs are, probably because owl eggs are covered during the day by the incubating female and the eggs are often concealed in tree hollows or deep grass. Owl eggs do not need to be camouflaged and white may help to make them visible in hollows. Compared to other birds' eggs, owls eggs are comparatively round, a shape that occupies less space for their volume than do other configurations, as one would expect for hollow-nesting species. Owls nesting on flat areas, such as Snowy Owls or Short-eared Owls, lay more oval-shaped eggs (Johnsgard 2002). *Tyto* owls are said to begin incubation from the first egg and be less seasonal than *Ninox* owls that begin incubation with the second egg, if there is one. Consequently, nestling *Tyto* owls (such as Barn Owls) within a brood may be of several ages, but nestling *Ninox* owls in a brood will be closer to the same age.

In the days before egg-laying, females sit fluffed-up and looking ill. This is called 'egg-laying lethargy' (see Figure 27.1). This period of inactivity allows eggs to form and this appearance of 'lethargy' stops after eggs are laid.

The clutch size in Australian owls varies with each species, but the difference is not closely linked with the body mass of the owl (Table 27.1). The four

Figure 27.1 Females in egg-laying lethargy often sit at the nest or on a branch nearby, puffed up and looking ill.

Australian *Ninox* species show little variation in clutch size, though the two smallest, the Boobook and Barking Owl, show a slightly greater range in clutch size than do Powerful and Rufous Owls. The four *Tyto* species show more difference in both clutch size between breeding attempts for a given species and the range of clutch size between species. The forest species, Masked and Sooty Owls, have smaller clutch sizes and ranges than do open-area species, the Grass and Barn Owl. Within European *Strix* owls, the species with larger clutch sizes and ranges are said to be typical of *stenophagous* owls, that is, owls restricted to a small number of prey species, such as the Great Grey Owl. This contrasts with *euryphagous* species, owls that use a broad prey base, such as the Tawny Owl, which have smaller clutch sizes and ranges. This may hold slightly with Australian *Ninox*. Powerful Owls and Rufous Owls probably do have a narrower prey base than do Barking Owls and Southern Boobooks, but not as markedly as in European *Strix* owls. The two Australian forest *Tyto*, Masked and Sooty Owls, probably have narrower prey bases than do the open-area Grass and Barn Owls.

Table 27.1 Clutch size, main food and habitat types of Australian owls (after Kavanagh 2003)

Species	Female mass (g)	Clutch size		Main food types	Main habitat type
		mean	range		
Powerful Owl	1350	1.9	1–2	arboreal mammals, bats, birds	forest
Barking Owl	710	2.2	1–3	birds, insects, ground mammals	woodland
Rufous Owl	805	1.9	1–2	arboreal mammals, bats, birds	forest
Southern Boobook	315	2.5	1–4	birds, insects, ground mammals	forest and woodland
Sooty Owl	1250	1.5	1–2	arboreal and ground mammals	dense forest
Masked Owl	835	1.6	2–4	ground mammals	forest and woodland
Barn Owl	360	4.0	1–7	ground mammals	open woodland, grassland
Grass Owl	310	6.0	3–8	ground mammals	grassland and heath

However, there may be other factors at play in the variation in clutch size and range in Australian owls. Kavanagh (2002a) said that Powerful Owls took 62 per cent Ringtail Possums by number, and Debus *et al.* (2004) said that Barn Owls took 62 per cent House Mouse by number. Why then do these two species have such differences in clutch size and range? Probably stenophagous owls rely on a small number of highly fluctuating key species, whereas euryphagous species do not rely on such fluctuating prey species. Key species such as the House Mouse or Long-haired Rat fluctuate at much greater magnitudes than do Ringtail Possums. The key factor may not be the number of key prey species that an owl takes, but how much the key prey species fluctuates or cycles.

Powerful Owls may be more *k-selected* than are species such as Barn Owls, which are more *r-selected*. The k-selected species are long-lived, slow reproducing and their populations stay fairly constant near the carrying capacity (K) of relatively stable environments. One might expect this in south-eastern Australian forests. In contrast, Barn Owls, more r-selected, have potentially larger ranges in clutch and brood sizes, reproductive rates and dispersal systems that allow for rapid changes in population sizes and the ability to change locations. Barn Owls would adapt to more unpredictable environments and carrying capacities, such as Australian grasslands and the arid inland.

Unlike many raptors, male owls do not incubate, at least in the species so far studied. During incubation the male normally feeds the female, or guards territory borders. However, males do defend nests, and incubating females at Southern Boobook nests we observed hunted right through the incubation period, sometimes leaving their eggs for up to an hour when outside temperatures were close to 5°C. Apparently, eggs are protected inside tree hollows and the wood lining; wood chips may insulate eggs and nestlings. Eggs can withstand considerable cooling at the early stages and remain viable (Les Boyd personal communication). During our watches on Black Mountain, males often took up to an hour to find, capture and deliver vertebrate prey, and females hunted insects closer to the nest while males were gone.

Females of all raptor species sit tight when the eggs are close to hatching. Even large *Aquila* eagles, the wariest of birds of prey, will sit tight when humans disturb the nest, and many female Collared Sparrowhawks or Southern Boobooks refuse to leave hatching eggs at all. They can be lifted off the clutch, banded and will resume incubating or brooding when the bander releases them. However, we do not advise this level of disturbance.

28
Pair behaviour – prey deliveries

Owls, like other birds of prey, capture most prey with their talons and deliver this prey to nidicolous young, that is, altricial young that stay in the nest until branching or fledging. Newton (1979) described three overlapping phases of female behaviour typical for breeding raptors with nestlings:

1) almost continuous brooding or shading before the young have developed their own thermoregulation;
2) presence near the nest for feeding and defending the largely helpless young; and
3) attendance with flexible periods of time spent distant from older nestlings or fledged broods.

Attendance patterns during the third phase varied greatly, and Newton suggested that the variation depends largely upon the hunting success of the male. Females seemed to leave broods only when the males' hunting efforts were insufficient and chicks required additional food. Researchers develop experiments to test such hypotheses around food shortage, by artificially increasing the number of nestlings (adding chicks from one nest to another) or decreasing brood size (removing chicks from a brood so adults do not have to hunt for them). Experiments largely confirm Newton's hypothesis. Female flight time increased when broods were increased in Eurasian Kestrels, and female hunting ceased almost completely in reduced broods (Dijkstra *et al.* 1990). Female American Kestrels spent more time hunting when broods were enlarged in a poor vole year, but there was no clear pattern in a year of higher vole abundance (Gard and Bird

1990). In one of the most elegant studies, four nests of Great Horned Owls were manipulated that originally held two young each during high densities of cyclic Snowshoe Hare populations in south-western Yukon, Canada (Rohner and Smith 1996). Broods of two nestlings older than 35 days were temporarily enlarged by one young, and then by two young. Radio-tagged parent females with enlarged broods did not respond to the addition of one young, but did respond to the addition of two young by moving further from their nests at night, apparently increasing their hunting effort; they spent less time near the nest during the day. When extra food was supplied to the broods enlarged by two, the females' behaviour returned to normal, and they stayed closer to the nest and guarded young.

Because Australian owl species are mostly nocturnal, it is difficult to observe feeding rates. Large owls such as male Powerful and Rufous Owls forage for long periods and often capture and deliver only one item per night. However, female *Ninox* with young sometimes capture moths, beetles, centipedes and other invertebrates close to the nest while males are off hunting. These invertebrates could be quantified by observers and would change the statistics around prey deliveries for some *Ninox* owls. Invertebrate prey would normally be quantified in studies of smaller raptors or corvids, but are sometimes left out of studies of larger *Ninox*.

There are no quantified data for prey deliveries to nests of any Australian species, except for a small number of observations at a Barking Owl nest (Barnes *et al.* 2005), and few data for nocturnal owls elsewhere. Delaney *et al.* (1999) found that prey deliveries for the Mexican subspecies of Spotted Owl were 2.68 items per 24-hour period during incubation, 4.10 during brooding, and 4.51 during the nestling phase. These owls delivered 0.37 prey items per hour at night and 0.03 per hour during the day. There are some limited data for easier to watch diurnal owl species such as the Northern Hawk-Owl, an owl similar in mass to the Southern Boobook, but more of a small mammal specialist. A pair in Sweden fed young once per hour during daylight; after 17:00 feeding decreased (Mikkola 1972). In Finland, they delivered prey 0.8 to 2.5 times per hour, and fed a brood of 1 nestling every 30 to 40 minutes at night, and every 95 minutes during the day (Leinonen 1978; Simon and Simon 1980). In Manitoba, parents averaged 2.1 deliveries per hour at one nest after the young had fledged (Lang *et al.* 1991). The Long-eared Owl is another mammal specialist about the same mass as the Southern Boobook and, like the Boobook, but unlike the Northern Hawk-Owl, it often hunts nocturnally. At two nests in Utah during the first half of the evening males delivered 1.8 to 2.5 prey items per night while females were incubating, and 3.5 to 4 items per night during the nestling period (DeLong 1982). In a study in Idaho, male Long-eared Owls fed branching fledglings 2.5 times as often as females did (Ulmschneider 1990).

The Eastern Screech Owl is about two-thirds the mass of a Southern Boobook and has a similar breeding ecology and is nocturnal. Frederick Gehlbach studied prey deliveries to nestlings at 19 suburban nests in Texas. The deliveries each night were distinctly bimodal, with the major peak at dusk; adults delivered about 76 per cent invertebrates. The owls made about 2.4 deliveries per hour in the early brooding phase to 5.1 deliveries per hour when nestlings were three weeks old. Gehlbach believed that fewer prey items were brought early in the nestling phase because it was cooler, so, instead of invertebrates, males brought larger vertebrate prey that could be stored. Later in the season, as insects and reptiles became available, adults made more prey deliveries with more invertebrate prey. In south central Idaho the small, insectivorous Flammulated Owl made about 8.45 deliveries per hour, mainly lepidopteran and orthopteran prey, with no vertebrates. In Colorado Linkhart and Reynolds (1987) found they made 9.9 deliveries per hour.

All of these studies reflect the division of labour between the sexes in birds of prey. Males deliver most prey during the incubation and early nestling phase of breeding, after which females began to hunt for nestlings.

To study prey delivery rates in Southern Boobooks we walked into woodland at sundown, several nights each week if possible, stood under the nest or daytime roost of one of the pairs of owls, and watched them from just before they left their roost or nest for one or two hours. If you slowly acclimatise owls to your presence, they eventually trust you and hunt and deliver prey as normal. With binoculars and a torch, or sometimes a digital camera, you can record the type of prey and the sex of the adult who delivered it. Better-funded research fixes cameras at a number of nests to record this sort of data but standing close to a nest gives flexibility; you can observe hunting beyond the perimeter of the nests as well as deliveries to the nest.

We observed prey deliveries at eight Southern Boobook nests over 183 nights between 1993 and 2008 on three adjacent nesting territories. In 1993 Jeanine Randall recorded prey deliveries for the White pair (Figures 28.1 and 28.2). We noted all prey deliveries and categorised these as male or female, in the first half of the nestling phase (when males should do all the hunting), second half of the nestling phase and fledgling phase (when both adults should be hunting).

Females usually did not travel far, but males could be gone for an hour. As the nestling period is about six weeks, we called weeks 1 to 3 the first half of the nestling period, weeks 4 to 6 the second half. In weeks 1–3, females delivered significantly more prey than males (Table 28.1), partially because males delivered more vertebrate prey and these took longer to find (Table 28.2). In weeks 4–6, there was no significant difference between the number of prey items delivered by males and females. After the young fledged, we followed individuals after they left the roost or nest as closely as possible without disturbing them and calculated the frequency per hour of prey deliveries to fledglings. In the fledging period, there was no significant difference between the number of prey items delivered by males and females.

Prey deliveries for owls in this study averaged 4.57 and 5.5 deliveries per hour for males and females respectively over the nestling and fledgling period

Figure 28.1 Male White delivering prey.

combined. These values were higher than those for some similar-mass owls such as Northern Hawk-Owls and Long-eared Owls cited above, but similar to the deliveries per hour for Texas Eastern Screech Owls which delivered 2.4 items per hour in the early brooding phase, and 5.1 per hour to older nestlings.

Newton's (1979) three overlapping phases of female behaviour typical for breeding raptors with nestlings: (1) almost continuous brooding, (2) presence near

Table 28.1 Mean number of prey deliveries per hour by male and female Southern Boobooks to nestlings and fledged young (per cent in parentheses)

	Nestling period						Fledgling period		Nestling + fledgling	
	1st half 0–3 wks (% male/female)		2nd half 4–6 wks (% male/female)		Nestling totals (% male/female)		Totals (% male/female)		Totals (% male/female)	
	m	f	m	f	m	f	m	f	m	f
Mean total deliveries/hr	1.6 (32)	3.44 (68)	5.11 (47)	5.84 (53)	3.9 (44)	5.01 (56)	5.29 (48)	5.75 (52)	4.57 (45)	5.5 (55)
Mean total deliveries/hr (male + female)	5.04		10.95		8.91		11.04		10.07	

Figure 28.2 Southern Boobook Male White arriving at the nest with prey.

the nest for feeding and defending and (3) flexible periods spent close to or distant from older nestlings or fledged broods, do not fit the patterns found with Boobooks in this study. In Table 28.1 the females made 56 per cent of deliveries during the nestling period, males made 44 per cent. Females made 52 per cent of deliveries during the fledgling period, males 48 per cent, not the 'normal' pattern of male/female behaviour set out in many studies, in that males did not provide all food during the first half of the nestling period. In fact, during incubation, the female Boobooks hunted on their own on 52 of 103 observation nights, all insects in our observations (not counted as prey deliveries).

Female raptors need to guard the nest and young, but some female Boobooks could guard and hunt at the same time, if prey was close. Do any other raptors do

Table 28.2 Number of vertebrate/invertebrate prey deliveries at 6 Southern Boobook nests, 103 observation nights during incubation and nestling stages by males and females (per cent in parentheses)

Breeding Stage	Incubation (%)		Nestling (%)		Fledging (%)	
	Vert	Invert	Vert	Invert	Vert	Invert
Male	15 (14)	93 (86)	15 (3.4)	421 (96.6)	1 (0.3)	372 (99.7)
Female	–	–	7 (1.2)	576 (98.8)	3 (0.75)	397 (99.25)

this? Mississippi Kites *Ictinia mississippiensis* behave in a similar fashion. If there is little danger of predation, females can hunt within sight of the brood, and the brood is unlikely to suffer from hypothermia. The females of this kite hunt through the early nestling period, as Boobooks did.

Botelho *et al.* (1993) studied a Mississippi Kite nest in New Mexico and found that over the six weeks of the nestling period males delivered 4568 prey items, and females delivered 2924 items, but in the first week after hatching there was very little prey transferred by the male to the female. Prey was mainly Apache Cicadas *Tibicen sayi* and over the first two weeks after hatching prey items delivered were: week 1, male 514, female 520; week 2, male 735, female 900. Southern Boobooks and Mississippi Kites are about the same size and mass, and have the same direction and degree of dimorphism. During the nestling period, male Mississippi Kites and Boobooks tend to bring larger prey to the nest than do females because males hunt vertebrates away from the nest, while females hunt invertebrates close to the nest. So, in terms of biomass, males bring more prey than would be evidenced from only counting prey deliveries.

Results from this study may hold broadly for Australian *Ninox* with 'normal dimorphism', males larger than females. For example, female Rufous Owls hunt insects around the nest and males hunt through the night for large prey that can be as heavy as the male (Hollands 2008). The balance between guarding nestlings and eggs from predation and foraging is met in different ways, but direction of dimorphism does not seem to be a factor.

The nature of female guarding could also affect studies of prey delivery rates in raptors. We climbed some Boobook nests where females were regularly delivering prey and, after this, two female Boobooks stopped delivering prey if we were present. They watched us and gave Single Hoot calls instead of delivering prey. They were excluded from the study. Studies that examine prey deliveries at the same time researchers are visiting the nest to collect prey remains or experiment with young are likely biased against females hunting, and these results should be treated with caution. That some females *do not* hunt or contribute to foraging apparently increases the fitness of the brood, that is, guarding the young from predators or inclement weather increases the number of young fledged compared with bringing more food to those nestlings. Alternatively, females that avoid hunting increase their own long-term fitness – gain longevity compared with females that do hunt. However, it is not clear how females that avoid hunting benefit their mates. These factors may vary with species, prey type (vertebrate v. invertebrate) and circumstances. The pattern of prey deliveries in Southern Boobooks and Mississippi Kites may contrast with some other raptors; for example, male Prairie Falcons (*Falco mexicanus*) at some nests do virtually all of the hunting in the first half of the nestling period, perhaps because there is more of risk of predation or chilling to nestling Prairie Falcons than to hollow-nesting Southern

Boobooks or colonial-nesting Mississippi Kites. More important, Boobooks and Mississippi Kites can hunt and guard young at the same time because females are hunting abundant, easily caught insects rather than difficult vertebrates.

In our study, males would sometimes arrive with prey, call to the female, go into the hollow, fail to find her, then go looking for the female instead of giving the prey item to nestlings. We interpreted this as a male 'showing' his mate that he was providing, showing her that he had large prey for his time away, even if it was a large-looking moth. This might encourage her to stay during that breeding season and return the following year.

We suspect that the ratios of male/female hunting we saw in Boobooks were typical of the species and not necessarily linked to food shortage, as could be inferred from a reading of Newton (1979). Researchers have a narrow view of sex roles in owls. Females probably hunt more than has been previously reported, as long as their young are not at risk from predators or weather.

29
Fledging age

How old is an owl when it fledges? This should be a simple question; for example, how old is a young Southern Boobook when it leaves the nest (Figure 29.1)? It turns out that it is not such a simple question. We can start with David Fleay (1968) who first bred Boobooks in captivity. He, and Schodde and Mason (1980) and Hollands (1991) said that young Boobooks fledged (flew) at about six weeks old. Others – del Hoyo *et al.* (1999) argued that mainland Boobooks fledge at around five weeks, not six weeks as Fleay and others had claimed. Surely the later del Hoyo reference was more accurate, based on better science, more precise definitions and better measurement techniques? Not necessarily.

There is confusion in the literature around the definition of fledging. According to one standard text, *A Dictionary of Birds*, fledging means a bird leaving the nest when it is ready to fly and has completed its first complement of true feathers (Campbell and Lack 1985). However, owls and some other birds can leave the nest before they fly and before flight feathers (primaries and secondaries) and tail feathers are fully grown. If these birds have left the nest but are not fledged, what is this stage?

Owls, and some hawks and eagles, have young that 'branch' into foliage adjoining the nest. Likewise, owlets crawl out along branches flapping their wings to balance, using their claws and sometimes their beaks to move out to the ends of limbs, then they crawl back to the nest. They sometimes slip at this stage and fall to the ground, or land further down in an adjoining tree. Humans can pick them up and they end up in rehabilitation centres, where they should not be. Owlets can glide, but they cannot fly. They often try to get up off the ground by walking to a

Figure 29.1 Fledged juvenile Boobook (foreground) with its mother.

nearby tree and climbing up, using beak, talons and flapping wings to get as high as they can. Finders should pick up these young owls and, when it is getting dark, place them in a tree close to where they were found; the parents will find them.

Juvenile Boobooks that have left the nest are called 'branchers', and owl researchers such as Johnsgard (2002), give different ages for 'branching' in contrast to 'fledging'. For Barred Owls in North America Johnsgard said, 'When about four or five weeks old they leave the nest and clamber about on nearby branches, but do not fledge until they are about six weeks old'. He defines 'branching' as, 'A term applied to the behaviour of unfledged owlets that leave the nest and begin clambering about on the branches near the nest site'.

Ian Newton (1986) differentiates 'branching' and 'fledging' in Eurasian Sparrowhawks. Nestling males move onto branches near the nest at 19 days old, the larger females at 21 days old. Males make their first flight at 26 days, females at 30 days.

We found that Boobooks fledged later if the nest hollow was in a smooth-barked tree like a Scribbly Gum (difficult for owlets to climb up), and if there were no branches or ledges nearby for the owls to crawl to, where they can walk, receive food, then move back into the nest for the day. Older nestlings with no ledge or close branch sometimes fixate on a branch a metre or two from the nest, study it

Table 29.1 Fledging ages for Australian owls cited in Higgins (1999)

Ninox	
Tasmanian Boobook	5 to 6 weeks
Southern Boobook	5 to 6 weeks
Barking Owl	35 to 36 days
Rufous Owl	7 to 9 weeks
Powerful Owl	7 to 9 weeks
Christmas Island Owl	68–77 days (from assumed date of laying)
Tyto	
Sooty Ow	about 3 months
Lesser Sooty Owl	about 3 months
Masked Cwl	10 to 12 weeks
Tasmanian Masked Owl	10 to 12 weeks
Eastern Barn Owl	50–55 days
Eastern Grass Owl	4 to 5 weeks

for several minutes while perched in the nest entrance, then suddenly fly to the branch and land clumsily.

The difference then, given for fledging age of Southern Boobooks in the literature, five weeks versus six weeks, could reflect variation within the species or between nest sites, but, more likely, the differences between definitions of branching and fledging. Fleay describes the juveniles as flying across the cage, while the source used by del Hoyo *et al.* (1999) said that owlets left the nest with flight feathers about 70 per cent grown. This would be 'branching'. Higgins (1999) defined fledging as when a bird first leaves the nest and fledgling as a bird that has left the nest at least once. This, we think, is misleading because it ignores the important 'branching' stage of raptors. Many owlets crawl out to the lip of the nest hollow, crawl around to the side of the trunk if possible, then return, well before they first fly or glide. Species such as Grass Owls leave the nest well before they can fly and run along the ground. Consequently, the fledging ages for Australian owls in Higgins (1999) (Table 29.1) should be treated with caution.

In short, Southern Boobooks fledge at around 5 to 6 weeks, but this can vary with definition of fledging and type of nest. Other Australian species will show at least as much variation and need further study.

30
Female desertion

Most ornithologists know about the separation of home duties that characterises birds of prey during the breeding season. The larger female does most or all of the incubating of eggs and brooding of small nestlings, while the smaller male does most of the hunting. With owls, in contrast to falcons, females do *all* of the incubating while many males travel long distances in search of food. Females stay back and defend, often fiercely, the nest, eggs and young against predators, including humans.

What is less known about some raptors, like Screech Owls and Cooper's Hawks in North America, is that some females show an even more lopsided partitioning of duties after the young fledge. After the young fledge, but are still dependent on their parents for food and protection, many females leave. Our female Boobooks refused to feed or defend their fledged young, even when these young sidled up to them along a branch and begged. Females abandoned all duties to their mates. Ornithologists call this 'Female Desertion'.

In the first few weeks after fledging, the mother Boobook often sat with her young and the adult male sat in a separate tree, maybe 50 m away. At night, on partly feathered wings, the young struggled to fly and gave persistent cricket-like trills as they moved through the forest, like children blowing tiny whistles in the dark. They followed their parents from tree to tree and the female defended them from ravens, kookaburras, possums and from us. She sometimes gave a loud Single Hoot as she swooped past our heads. Because Southern Boobooks in this study used the Single Hoot call most commonly after the young fledged, we heard it most often in December and January (Figure 19.4 in Chapter 19). Females used it just after the young fledged, then males used it when they took over parenting responsibilities, after the females 'turned off' or left the breeding area.

We studied post-fledging behaviour in three pairs of Southern Boobooks. Table 30.1 shows the number of days after fledging that females stopped feeding their young and left the brood, and also when the young became independent.

Different females stopped caring for their young in different ways. In one variation, males moved with their young to different parts of the territory away from the family's nest and 'camped out' during the day. As the male and his fledglings travelled further from the nest, feeding as they went, the female stayed behind. She did not feed or defend her young because they were not near the nest; they had left her.

In a second variation, some families, including females, roosted during the day close to the nest. After two to four weeks, the females 'lost interest', stopped feeding and defending their young. Oddly, the family might roost together during the day, so researchers monitoring the daily roosting behaviour of these owls might not see female desertion. The researcher might assume she was feeding and caring for her young. But this was not the case – males and females in this study were marked and followed after dark. Maybe some families roosted together during the day because the fledglings and the male sought out the female at dawn and roosted next to her, even though she ignored them during the night.

During this 'desertion' period females still showed affection to their mates, billing and mutually preening with him (Figure 30.1). But the adult male assumed total responsibility for feeding and defending the fledged young. After females left, males juggled three tasks – feeding their young, defending their young and defending territory borders. This put males under increasingly aggressive demands from their large young; aggressive female fledglings weighed more than their fathers. Fledglings attacked the male and took food from him.

A third way that females deserted was through short migrations. They left their families in January or February, sometimes only two weeks after the young fledged, and travelled to one of Canberra's suburbs, maybe 10 km away. They set

Table 30.1 Number of days after young fledged that (a) female stopped feeding young, (b) female dispersed, (c) young became independent, (d) young were fed solely by the male: *Pair # 1 1996–1997*, 30 nights x 1 hour of observation between 3 October 1996 and 3 February 1997, two young; *Pair # 2 1994–1995*, 59 nights × 1.5 hrs of observation between 20 September 1994 and 21 February 1995, three young; *Pair # 3 1993–1994*, 25 nights × 1 hour of observation between 6 October 1993 and 6 March 1994, one young

Pair	(a)	(b)	(c)	(d)
#1	23	34	48	25
#2	25	25	42	17
#3	14	30	43	29

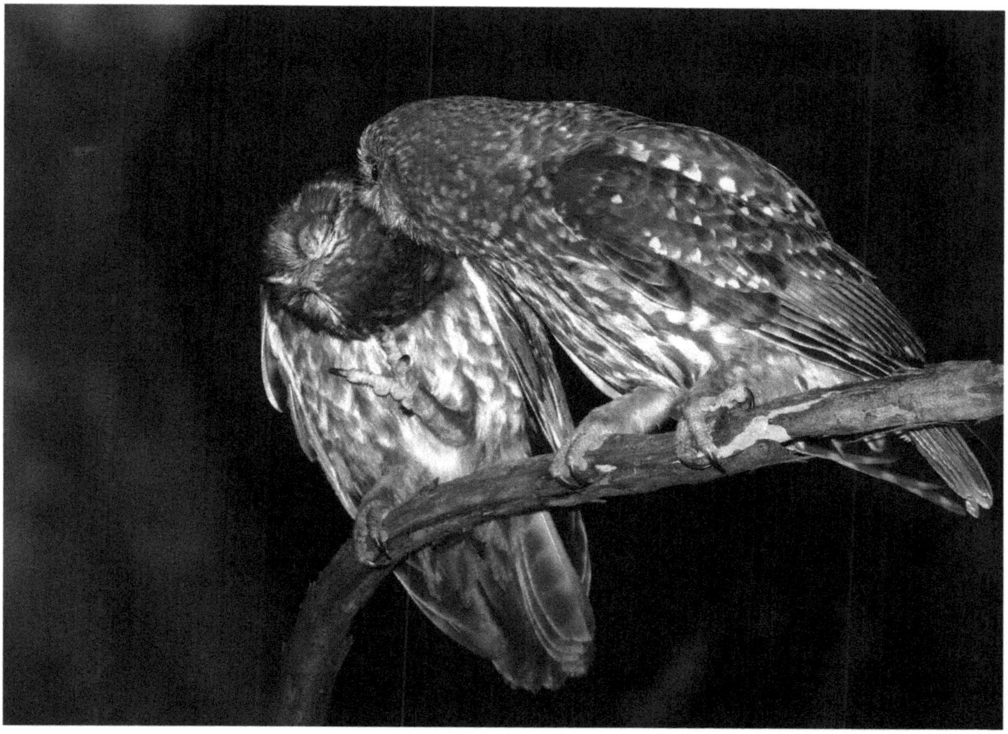

Figure 30.1 Some deserting female Boobooks ignore their young but still show affection to their mates.

up a winter home range and stayed until spring. One female winter home range was about 100 ha (Figure 8.2).

After his female deserted in 1993, White moved to his favourite roost on the southern border next to Green's nest and his young followed. Each night White Duelled with Green along this border, where Green had a single, newly hatched nestling in a hollow. White and his fledglings hunted insects and 'camped out' just over the border in a Native Cherry or eucalypt tree by day.

One night White's fledglings crossed into Green's territory, trilling and calling for food near Green's nest. They did not seem to recognise the border, or discriminate between adults. White did not follow, nor did he call to them. They begged food from Male Green, but Male and Female Green ignored them. Eventually the fledglings strayed back over the border and joined their father.

At first we thought White and his fledglings had travelled to Green's nest because the area contained abundant food, and White was defending this food source through Duelling. But later we believed that White was keeping an eye on Green, at a point along the border where Male Green had been trying to wedge into White's territory. White was determined to drive him back and he had to take his young to the fight, multi-tasking. White had triple duties – protecting his young, feeding his young and keeping his territory border in place.

Communication between adult males was complex, and could be explained by long-term investment in a territory, the defence of a perimeter. But did this matter to females? White's female stayed behind at the nest when he travelled over the hill, and female Green overheard or witnessed the Duelling between Green and White that summer. In the following year she left Green and mated with White, and stayed with him for six years before a cat killed her on her winter territory.

31
Fledgling behaviour

Fledging behaviour is poorly understood in Australian owls. The period between a juvenile's first flight from the nest and its dispersal from the breeding territory has received little attention. It is a crucial period for young owls because they need to practise hunting. If juveniles fail to acquire hunting skills during this period, they may starve, even in the face of abundant food. Most do acquire hunting skills, but they need to acquire them before winter cold and food shortage make life harder, at least in south-eastern Australia.

The values for fledging to independence given for Australian owls in Higgins (1999) are given in Table 31.1.

The most obvious behaviour in fledged owlets is food-begging. Most species use whizzing, churring, rasping, or hissing calls. Boobooks use a cricket-like Trill call and they do this persistently through the night. The Trill call is also a Contact call, allowing the young owls to find each other in the night and stay together. Young *Ninox* follow adults through the forest, or stay put and wait for adults to deliver food. During the day they may roost back inside the nest hollow, in a different hollow, or with all or part of the family in a eucalypt, exotic or Native Cherry tree.

Occasionally, fledgling Boobooks crossed territorial borders and begged from neighbouring adults. In these instances during the first 10 years of the study, the neighbouring parents ignored the intruding juveniles and pulled back their faces if food-begging juveniles perched close and touched beaks with them. Adults flew away if the neighbouring young persisted, and we did not see parents cross borders to retrieve strayed fledglings. The young seemed to treat all parents, even

Table 31.1 Days from fledging to independence Australian owls cited in Higgins (1999)

Ninox	
Tasmanian Boobook	6 to 7 weeks
Southern Boobook	6 to 7 weeks
Barking Owl	3 to 4 months
Rufous Owl	several months
Powerful Owl	150 to 240 days
Christmas Island Hawk-Owl	at least 2.5 months
Tyto	
Sooty Owl	3 to 5 months
Lesser Sooty Owl	several weeks +
Masked Owl	10 to 12 weeks
Tasmanian Masked Owl	10 to 12 weeks
Eastern Barn Owl	12 weeks
Eastern Grass Owl	not known

neighbouring parents, as food providers. They flew clumsily up to a neighbouring adult, sidled along a branch and trilled for food. Neighbouring adults did not harm intruding young, even if the adult was at war with the juvenile's parents. Neighbouring adults never attacked these confused young or drove them back over the border. The juveniles we observed found their way back home without help. One juvenile crossed a border, but did not find its way back before daylight. It roosted all day with a pair of adults that had no young. On nightfall he trilled for food and followed the surrogate parents around in the dark. They ditched him, flew out of sight. From these incidents we concluded that parents recognised their own young, probably by their Trill calls, and parents would not feed neighbouring young that strayed. This gave the impression that adults could discriminate between the vocalisations of their own fledged young and the vocalisations of neighbour's fledged young, in contrast to raptors like Peregrine Falcons *Falco peregrinus* that readily feed young other than their own. You can transplant nestling eagles, hawks and falcons into different nests and the unrelated adults will raise them.

However, Ed McNabb (1994) successfully fostered a fledgling Powerful Owl with a pair of owls that had a single similar-aged young, so some *Ninox* owls will apparently feed young other than their own under certain circumstances.

The case study below tested some of these assertions, and showed the complexities of the Boobook post-fledging system.

Between 1993 and 2002 there was a high frequency of Duelling between neighbouring males (Chapter 35), but the neighbouring males Orange and White never Duelled with each other, only with other males on their outside borders. We

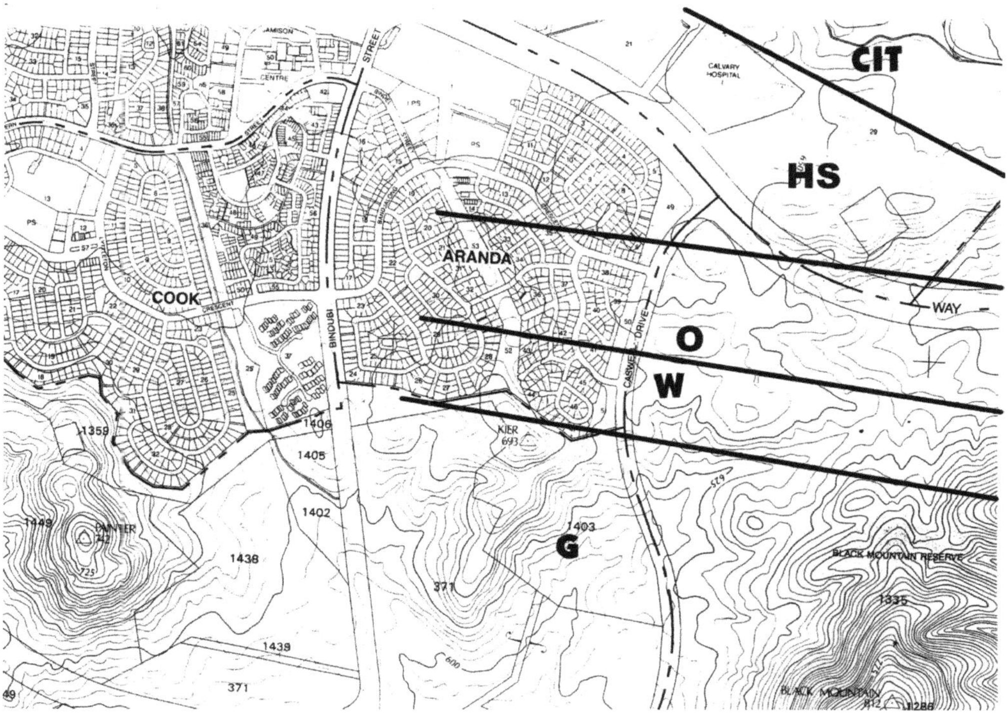

Figure 31.1 Nest locations and borders. By 2002 Males White (W) and Orange (O) nested close to each other, apparently because of conflict on their borders with Males Green (G) and Hospital (HS). CIT = Canberra Institute of Technology.

called the Orange/White border a 'peaceful' border, because there was no Duelling along it. We called the White and Orange outside borders 'hostile' because the owls Duelled with neighbours along these borders, especially the males Hospital and Green. Maybe White and Orange shared a long-standing border that became peaceful over time as they got used to each other, or maybe they were blood-kin (see Figure 31.1). They got along, but stayed out of each other's territories.

Boobooks in Aranda Bushland and on Black Mountain often nested close to territory borders, not in the centres of territories, even if neighbours were enemies. Though territories were about 1 km^2, the nests of both friendly and enemy neighbours could be much closer. In the past, White and Orange had nested close together, but their young had fledged two or three weeks apart. The Orange young were usually older and they sometimes met and mingled with the White brood, but separated during the day. Broods returned to their own territories and parents.

In 2002 White and Orange hatched three young each in Scribbly and Brittle Gum hollows about 200 m apart. Apparently they were pushed together by pressure from males on their outside 'hostile' borders.

On 12 December the Orange parents fledged one young and fed it in Stringybarks near the nest. By 19 December the two other Orange juveniles were out. The White pair still had three young inside their nest hollow, 200 m away.

A White juvenile fledged on 20 December. We trapped, banded and radio-tagged it so we could follow it and watch it learn to hunt, gain independence and leave the territory. A second White juvenile fledged on 23 December. Everything looked rosy. All the young followed their parents from tree to tree, and the parents caught insects in mid-air, scraped them from eucalypt leaves or caught birds and bats.

On Christmas day there was a problem. The two White juveniles had disappeared. The White parents sat in a Stringybark tree alone. We assumed the juveniles had died; some occasionally perish from accidents or predators. But we could hear a radio signal. We followed the signal over the hill through trees into the territory of their northern neighbour, Orange, half expecting to find the juveniles dead on the forest floor. Both of the missing young were in a tree with the Orange parents, alongside the Orange brood of three. Seven owls roosted together.

This troubled us. We hoped the owls would sort things out during the coming night, as juveniles had in the past. We knew parents could recognise their own young, so the Orange parents would never feed the two strayed juveniles. They did not know how to hunt and there was not enough food on a single territory to feed this many owls. The two young should get hungry and go home.

In the first hour after dusk, this was the case. The Orange parents ignored the frantic trilling of White's young, though all five young persisted.

We followed the five young around on Christmas night and the Orange parents fed their own fledglings. By the end of our watch they had not fed White's juveniles and White's juveniles had not moved back home (Figure 31.2). Instead, they travelled downhill away from their own nest and away from their own parents, following the Orange family. At the end of the observation period, White's juveniles sat alone in a Brittle Gum, fluffed up and hungry.

Meanwhile, the White parents treated the Orange/White border like a stone barrier. They stayed about 50 to 100 m uphill and would not cross, feed or retrieve their own young. At this point, it looked bleak. The two White young would starve.

The following day, 26 December, all seven owls roosted by day in the same tree. We did not know if the two visitors had been fed; as the sky darkened the White juveniles left with the others. We did not see the White juveniles drop any castings, the packages of undigested insect legs, wings or vertebrate bones, fur and feathers. During our one-hour watch, all five young begged for food, but the Orange parents only fed their own young. At one point, the family wandered uphill towards the White border and we had a glimmer of hope. Maybe the White parents would cross over and retrieve their young. They did not, as if it was illegal to violate the border even with their young in peril, and even though White and Orange males did not appear to be 'enemies'.

Figure 31.2 Fledged juvenile Boobooks generally roost together, sometimes with neighbour fledglings.

That night the seven owls wandered back downhill away from the White parents. By the end of the hour the two visiting juveniles had not been fed. Social conventions normally gave order and predictability to owl society in the forest and safety for the young, but these conventions now worked against them. The two White owlets were confused, the neighbours would not feed them. The White parents would not rescue them and they would starve.

On 27 December there was no indication that White's young had been fed, and they acted hungry. More troubling, at dusk they flew downhill – away from the White border and their own parents. They ended up some 350 m downhill and alone. They were doomed.

But around 9:30 pm as if the frantic trilling over the past three nights had become too much, the Orange male arrived with a Gould's Wattled Bat and fed it to the White juveniles (Figures 31.3 and 31.4). They squabbled for a moment, then one took it and devoured it. We recorded this on film.

A rule had been broken. An adult had fed someone else's young.

On 30 December the Orange adults fed the White juveniles at least three times during our watch. Maybe the White juveniles would survive.

Figure 31.3 Male Orange (bottom left hand corner) arrives with a Gould's Wattle bat for the two 'lost' White juveniles.

Figure 31.4 White juvenile consuming bat brought in by the Orange male.

On New Year's Day 2003 there was a new problem. The last and youngest White juvenile had fledged and crossed the border. Six juveniles, three Orange and three White, now roosted with the Orange adults. As the sky darkened, this new owlet started to move with the others but had trouble keeping up. He was a poor flier and very young. Normal brood size is about 2.6 young per pair, three is common, occasionally there are four, but six would be too many to feed. The Orange adults tried.

On 2 January the White adults still roosted alone on their territory, 100 m uphill from the brood that had abandoned them. We talked about whether to interfere. This seemed to be a simple glitch in the breeding system, though a cruel one. Maybe we should let it run its course. The young owls were attracted to the neighbours' adolescent trilling more than to their own parents. But we had trouble watching the smallest owl starve.

We trapped him close to the border, banded, radio-tagged him and carried him back to the White pair at 22:45 hours. The White male came in, purred quietly, gave soft Contact Boobook calls, and after 15 minutes we left them sitting together.

The following day the youngest White fledgling was back with the Orange juveniles. He had crossed the border again and all six juveniles sat with the Orange adults.

The White adults stayed 100 m uphill, close but refusing to cross. When the White female left the roost for the evening, the White male followed her, sat next to her and face-preened her, as if they had no young, no responsibilities. Neither of them flew towards the border and their own young.

The six White and Orange juveniles left the roost with the Orange adults. The smallest had not been fed and, because he was so young, we expected him to weaken and die. The Orange fledglings were hunting cicadas at around 20:42 hours, but we did not see the adults deliver any prey. At this point, we had not seen the smallest White fledgling get any food.

Three days later on 6 January the two oldest White juveniles were hunting. It was clumsy, but they caught a few insects in mid-air. The youngest White fledgling did not hunt; he was too small. We had not seen him fed for three nights. But on 7 January, we saw him cough up a small casting at the day roost, containing insects. Somehow he was getting a little food, though we did not know how.

That night he would fly up to the other juveniles just after the Orange parents fed them, but did not seem to get any food from them. However, on 20 January we saw him fly up and take an insect from an Orange juvenile. The older fledgling released the insect, treated him like a demanding little brother. There was no antagonism. This seemed to be one source of food for the youngest and smallest White juvenile.

We saw the Orange adults feed their own young between 20 and 26 January, but never saw them feed White's young. Maybe White's young were catching prey at a younger age than normal. The Orange adults did not feed them enough, but all six fledglings were sharing prey – a big chummy juvenile gang.

On 23 January the Orange female deserted the young, as females normally do (see Chapter 30). She left all six juveniles 37 days after her own young had fledged and we did not see her again. The six juveniles were alone now with the Orange male.

On 31 January the three Orange juveniles still roosted together, but the three White juveniles had disappeared. We hiked through forest trying to find them with the radio-receiver, but had no luck. They were not uphill with the White parents, and they were not at the east end of the territory with the Orange male.

We climbed into the car and drove north, stopping every 500 m to test patches of woods with the antenna, listening for the signal. It took an hour, but we found two of them. White's oldest and youngest were about 3 km north of the Orange nest, across a major roadway in the suburb of Bruce. This new location was outside the Orange territory, and even further away from White's territory, but inside the territory of a third pair. The two White juveniles chirruped and hunted together and did not mix with the territory owners or their young. The brother and sister roosted together during the day in a bushy tree next to a walking path and nobody noticed them. They had dispersed 41 and 43 days after fledging, which is a little early, but close to normal. Maybe White's young became independent at an earlier age than the Orange young because the Orange adults had not fed them. We never found the third White (radio-tagged) fledgling. Apparently she had travelled in another direction beyond the range of our receiver.

On 8 February 2003 the Orange juveniles left the Orange territory, 52 and 53 days after fledging, a normal age for dispersal. The two White juveniles still roosted and hunted 3 km north in the suburb of Bruce. It seemed now that this youngest juvenile, who we thought would starve, would now make it. He was living on a quiet side street with his sister, and hunting well. On 27 February, they both disappeared, as is normal.

32
Snowy Owls

Vancouver is cradled between forested mountains in south-east Canada. Mist hangs over the city for much of the winter. On some days it breaks, giving residents a view of snowy mountains around town, a gift for enduring the head-down, drizzling rain. Canadians living away from Vancouver call it 'the Tropics' because it hovers around 0°C through winter, warm compared to the rest of Canada. It is not an environment for Snowy Owls, but quite a few lived south of town in 1996–1997.

A few days before New Year's Eve in 1996, Vancouver was ice-covered and paralysed, knee-deep in snow dumped by a series of cold fronts. The city council had spent its snow-clearing budget and winter was only half over. But cold did not force Snowy Owls south into Vancouver that winter, lack of food did.

The Snowy Owl *Bubo scandiacus, Bubo* meaning a hooting or horned owl and *scandiacus* a Latinisation of Scandinavian, was previously known as *Nyctea scandiaca*, *Nyctea* from the Greek *nycteus*, 'nocturnal' but Michael Wink showed they were genetically close to the *Bubo* eagle owls.

Their breeding distribution looks like chocolate poured over the roof of the world, along the north coast of Alaska, Canada and its Arctic islands, northern Norway and Sweden and northern Russia, with some dribbling a little way south. This treeless tundra has 24-hour daylight in midsummer and nearly 24-hour dark in midwinter. Many Snowy Owls stay north in midwinter, but in 1996–1997 many irrupted southwards.

With dense plumage, Snowy Owls are designed to survive winter ice storms. Even their hollow feather cells trap air. Snowy Owls are the heaviest owls in North

America. At about 1650 g for males and 2000 g for females (Earhart and Johnson 1970), they are heavier than the average Great Horned Owl and heavier than the Australian Powerful Owl. The Eurasian Eagle Owl and Blakiston's Fish Owl are the two heaviest species in the world.

Typical breeding habitat for the Snowy Owl is open hummocky or rolling ground, not unlike the top of Mount Kosciuszko. The nest is a shallow depression on tundra; both male and female scrape and gyrate a hollow into the vegetation.

In Finland, they lay five to 14 eggs, depending on peak or low rodent years, a large number of eggs for a raptor. They lay an egg every two to five days and, as long as the egg is not frozen, it can lie on the tundra for many days unincubated, and the embryo will not die. The embryo starts to grow when the female incubates. Many bird species delay incubation until most or all of the clutch is laid, so all the chicks hatch at around the same time. Snowy Owls do not do this. The female starts the 32 days of incubation, give or take a few days, immediately after the first egg is laid, so nestlings hatch at different times. Biologists find old nestlings alongside very small ones. This is said to be a solution to fluctuating food supply, but I suspect it is also a hedge against unpredictable Arctic weather. On some days blizzards or fog can stop the male from hunting or discourage prey from coming out into the open. In years of plentiful and available food, all the nestlings might survive and fledge, but in years of inclement weather or crashing prey, the younger, weaker nestlings die and may be eaten by their siblings.

Snowy Owls are big, obvious, white owls, and easy to spot by day. Their tundra breeding sites can be in low valleys or on elevated tablelands with abundant prey and a good view over surrounding countryside, nothing like the fenced grassy pasturelands around Edison south of Vancouver where Snowy Owls concentrated that winter, except for the flat terrain and abundant prey.

Raptors that catch small mammals are generally slow-flying species; they don't need to be fast. But Snowy Owls are fast. They fly with a rowing motion with deep beats on tapering wings. Ducks are amongst the fastest birds, and Snowy Owls around Edison stayed on their tails, matching them twist for twist as the ducks tried to escape.

On their northern breeding grounds, Snowy Owls prey mainly on small mammals, especially lemmings (in the genus *Lemmus* and *Dicrostonyx*). Lemmings are fat, snubby-nosed, short-tailed little mammals that live mostly in the far north, so they closely match the distribution of Snowy Owls. Their long soft fur nearly conceals their thin round ears.

We describe a person headed for certain doom as 'lemming-like', because Disney films show lemmings committing mass suicide running over cliffs into the ocean. These films were faked. The film-makers apparently 'coaxed' or threw lemmings off cliffs to recreate a myth current at the time. Lemmings don't suicide en masse, or on their own. They experience peaks and troughs in numbers, and

Snowy Owls respond to these peaks and troughs by moving to ground that is swarming with lemmings and leaving ground that is not.

Not all Snowy Owls face gluts of prey or starvation because of high or low lemming numbers. Lemmings do peak in certain places, scrambling and scratching over the tundra looking for food and this makes them available to Snowy Owls. However, Mikkola (1983) showed that Snowy Owls eat more *Microtus* voles than lemmings; these small mammals too are famous for driving peaks and troughs in the cycles of raptors. But on the Shetland Islands, there are no lemmings or voles – there Snowy Owls prey mainly on European Rabbits. Meinertzhagen (1959) described one owl capturing and subduing a large Arctic Hare by grasping the hare with one foot and using the other foot as a brake in the snow. Snowy Owls also use their wings as brakes when they capture large animals. They stab talons from both feet into the victim, spread their wings like a parachute to catch the air and slow the victim over the snow.

Snowy Owls eat birds too, including Black Grouse, ptarmigan, Mallard, oystercatchers, lapwings and others. This partially explains their speed. Birds are faster and harder to catch than mammals and Mikkola notes that the flight of the Snowy Owl is more hawk-like than owl-like; 'remarkably fast', he says.

Those species of Northern Hemisphere hawk, falcon or owl that irrupt and sweep south in search of food tend to specialise on a limited number of terrestrial prey, such as voles, lemmings, Snowshoe Hares or grouse – animals that cycle. Most raptor species do not do this. Great Horned Owls at southern latitudes, say in Virginia, do not irrupt or move around very much. They stay in one place for their breeding lifetime. Even though Snowy Owls in the bad winter of 1996–97 left the winter-hammered landscape, swept down through Canada and arrived on rooftops, fence posts and bare ground around Edison, most raptors populations show an *equilibrium* over time where breeding numbers are relatively constant over many years.

Raptors that fluctuate in numbers (numerical response) or in breeding success (functional response) with prey numbers are sometimes said to characterise a 'Bottom-up' model of community organisation. For example, moving from the bottom trophic level to the top, the producers could be green plants with flowers and fruits, the herbivores lemmings and the carnivores Snowy Owls (Table 32.1). A 'trophic level' is a functional classification of the different organisms in a community according to these organisms' feeding relationships.

A theoretical 'Bottom-up' system in Australia could have producers as green plants with flowers and fruits, herbivores as Ringtail Possums that eat flowers and fruits and carnivores as Powerful Owls that eat possums (Table 32.2).

These models are concerned with community organisation through the control of nutrients between adjacent trophic levels, the linkages between these levels. In a Bottom-up model, or in a Top-down model, one considers the interactions between

Table 32.1 Possible Bottom-up model – Northern Hemisphere

3rd trophic level	Carnivore	Snowy Owl
2nd trophic level	Herbivore	Lemming
1st trophic level	Producer	Sedges and berries

Table 32.2 Possible Bottom-up model – Australia

3rd trophic level	Carnivore	Powerful Owl
2nd trophic level	Herbivore	Ringtail Possum
1st trophic level	Producer	Flowers and fruit

adjacent trophic levels, for example, between green plants and possums or between possums and Powerful Owls. The Bottom-up model proposes that control is from the bottom up – nutrients control sedge and berry numbers, which control lemming numbers, which control Snowy Owl numbers. In contrast, a Top-down model proposes that Snowy Owls control lemming numbers, which control vegetation, which controls nutrients. In the Australian example predators, Powerful Owls, would control herbivorous possums, which would control tree fruit producers, which would control nutrients. The Top-down model has sometimes been termed a 'Cascade model' and predicts strong positive and negative interactions across all trophic levels (Krebs 2008). These models have been developed from Northern Hemisphere ecosystems; for example, wolves preying on deer, deer affecting tree seedlings and tree seedlings affecting nutrients. Removal of predators such as wolves from these ecosystems results in an explosion of deer numbers and suppression of seedlings so trees cannot regenerate, though nutrients can be independent of these relationships. However, systems with Australian owls differ from systems with predators such as wolves in North America. Powerful Owls can take a number of species besides Ringtail Possums, and they may be able to switch to other prey such as gliders, fruit bats or birds if possums decline. Furthermore, the eucalypts that possums and gliders feed on may not be negatively affected by arboreal marsupials. However, Kavanagh (1988) has argued that Powerful Owls do deplete numbers of Greater Gliders in certain areas, and then move from depleted areas to areas with abundant gliders and deplete these, a form of Top-down model. This finding has not been replicated and needs to be verified.

Why do different raptor species and populations react differently to food supply? In the next chapters we examine some predator–prey issues.

33

David Lack's theory of limiting factors

To understand breeding strategies, population declines, population increases or stability, one needs to understand *limiting factors*. The British thinkers Thomas Malthus and Charles Darwin noted that no population of animals goes on increasing without limit. Many factors can limit animal populations, and a population can be held down (limited) by more than one factor. Krebs (2001) said a factor was limiting if a change in this factor produced a change in the density of the animal's population. Food and disease are commonly recognised limiting factors – one would expect fewer Powerful Owls in areas with fewer possums or fruit bats than in areas with an abundance of these prey, evidence that food is limiting. If rabbit populations are low in one area where the myxoma virus was introduced, and high in an area without myxomatosis, this disease may be a limiting factor. If the Dingo *Canis lupus* is shot and trapped in one area, and fox *Vulpes vulpes* numbers increase, Dingos may be holding down fox numbers. Food and nest sites are said to be the main limiting factors for raptors (Newton 1979), and this can be tested experimentally. If nest boxes are added to an area and the number of breeding owls increases, then nest sites, more than food, are limiting.

David Lack, the prominent British ornithologist, was born in London in 1910 and was a teacher until 1940. He was interested in the ways animal populations are limited, and how their numbers stay roughly the same each year even though breeding pairs spawn many young.

Lack was educated at Gresham's School Norfolk, and at Cambridge. In 1940 he took a year off to study bird behaviour in the Galapagos Islands, and he worked on the development of radar during World War II. This study of radar led to a classic study of bird migration patterns. His most famous book, *Darwin's Finches*, was a landmark study, and other books like *The Natural Regulation of Animal Numbers* had great influence on the way modern ornithologists think about breeding birds. He was the leading British ornithologist of his time, and in 1945 became Director of the Edward Grey Institute of Field Ornithology at Oxford.

Lack wrote about stability, equilibrium. Egg-collectors in England knew they could return each year to a nest of eagles, falcons or owls and find them in the same patch because breeding numbers were stable. Collectors found a pair of birds in the same place each year and, barring habitat destruction, poisoning or mass persecution, they would find them in the same place next year. Though birds produce enough young to double or triple their numbers each year, their breeding numbers did not increase, they remained stable over time. Why did bird populations not explode?

Lack argued that each pair of birds produces as many young as possible given the amount of food available to them. After breeding in spring or early summer, food shortage kills off most of the young in autumn and winter. This brings the population back to the previous level. Two adult owls produce three young = five owls in spring/early summer. Some young die in the nest, and winter food shortage kills off two juveniles and one adult, leaving two owls in the following spring. In spring, owls spread themselves out over the landscape relative to the amount of prey, and again produce as many young as they can. Their clutch and brood sizes match the available food supply. They lay the number of eggs that will produce the most young they can raise that year, termed the *Lack clutch size.*

Regulating factors differ from limiting factors because they mediate the increases and decreases in populations. Populations reach an *equilibrium*, meaning they do not decline to extinction or increase to infinity. But the equilibrium level can change. It might settle at 10 individuals per hectare, or change and settle at six individuals per hectare. Population regulation implies some sort of negative feedback loop that intervenes between increasing or decreasing density of individuals and potential limiting factors such as food, predation, disease or territoriality.

These factors may operate differently when animals are at higher densities than when they are at lower densities, for example, more owls may live in an area in a year when rodent numbers are high than when rodent numbers are low, because pairs of owls defend smaller territories when food is abundant and defend larger territories when prey is scarce. *Density-dependence*, a term first coined by the entomologist H.S. Smith (1935), was one mechanism responsible for this. A *density-dependent* factor is a regulating factor dependent on the total number of individuals in an area. It is a controlling factor that reduces the 'extra' number of individuals, and keeps the numbers stable. For example, if death rates in owls increase as the density of owls increase, the death rate is density-dependent. If the number of young fledged falls with increasing density, then fledging rate is density-dependent. If three pairs of owls fledged three young each in a season (nine young total) then density-dependent factors (starvation, predators, lack of shelter) killed nine owls during the winter, bringing numbers back to three breeding pairs in spring, an equilibrium is set for that area at that time.

As mentioned above, the two most important limiting factors for raptors appear to be food and nest sites. There are not enough food or nest sites to go around, so bird numbers are limited *spatially*. According to this view, territorial behaviour does not limit the number of breeding owls – the amount of food or number of nest sites limit them. Each breeding pair gathers as much food as it can without wasting too much energy defending a large, inefficient territory, or wandering over too large a home range. Some owls do not find enough food and they die. Some owls do not find a nest site and they don't breed because of lack of food. Furthermore, one area holds 20 breeding pairs, while another holds 10 breeding pairs, because density-dependent limiting factors control the number of breeding birds. As suggested above, the New South Wales coast has more breeding pairs of Powerful Owls than the ACT has because there are more possums, gliders or bats per hectare on the New South Wales coast than in the ACT. Both ACT and New South Wales Powerful Owls are regulated by *density-dependent* factors, and this is expressed in pair densities. In the ACT Powerful Owls pairs spread themselves out and defend larger territories to get enough food.

These principles can also act temporally. A decrease in food supply in one season can result in fewer pairs breeding, or those that do breed have fewer young. The number of pairs in an area may drop from 20 pairs to 10 pairs because 10 pairs leave, or die, and the number of pairs may increase from 10 pairs back up to 20 pairs when food supply increases again.

Finding a statistical relationship between food supply and breeding success is the Holy Grail to raptor biologists. You increase your chances of publishing a paper if you can show some change in breeding success or failure related to changes in food supply in a year or cluster of years. Journal editors and referees like the notion, it is intuitively appealing, but in most ecosystems it is surprisingly hard to show. Some raptors live on *many* prey species, and some live on only a *few* prey species whose numbers follow a cycle of highs and lows, as with Snowy Owls and lemmings. But raptor experts often study owls, falcons or eagle species that kill and eat *many* species, and these raptors are unlikely to show any year-to-year change in clutch size, brood size or breeding success with changes in prey abundance, at least changes that can be measured. Ecologists argue that failure to find correlations between food and breeding success comes from the difficulty in measuring prey numbers. Changes are measurable if the raptor has one or two major prey species, for example, a species of rabbit, vole or lemming. The relationship between raptor breeding success and food supply is amplified. You can see the visible peaks and troughs in prey numbers from year to year, and clear changes in raptor breeding success from year to year.

However, even at northern latitudes where this holds true, it may not be that simple. The Norwegian raptor biologist Y. Hagen (1969) described *primary* prey species such as voles and hares, and *secondary* (scarcer) prey species such as grouse. Hagen showed that when mammal prey crashed, raptors switched to prey species such as grouse. Predator numbers and breeding success remained stable in the absence of primary mammal prey. Only after grouse declined, did raptors start to decline, and then eventually they switched back to increasing mammalian prey.

Great Horned Owls at high latitudes, for example in the Yukon Territory, respond to peak densities of Snowshoe Hares with all pairs breeding and fledging larger broods (Rohner and Smith 1996). However, when a raptor lives on *many* prey species, as does the ubiquitous Great Horned Owl at middle latitudes in North America, the temporal relationship between predator and prey is masked, or does not happen. If rabbits decline, Great Horned Owls simply switch to another prey species – pheasants, mice or squirrels – and keep breeding and producing the same number of eggs and young each year. There is no change in breeding success when rabbits decline. This worries some ecologists because, in a density-dependent system, there appears to be extra food that raptors ignore until they need it. If one or two prey species decline, they switch to the prey species they have previously ignored. If a preferred prey species does not decline, the raptors leave the 'ignored' species alone during the breeding or wintering season. This contradicts David Lack's notion that birds lay the number of eggs at the beginning of the season that will hatch and produce the largest number of young, given the available food. Instead, there may be more available food than the pair needs, and the pair does not use all of this food unless there is stress of some kind during the breeding cycle. This 'buffer' against food shortage is available to them when times are tough.

We know that breeding birds have 'extra' food available because of *brood manipulation* experiments where biologists place extra young into the nests of birds, then see if parents can raise these artificially enlarged broods. Often birds can raise an enlarged brood; the parent birds increase the amount of food brought to the young and successfully raise all of them. For example, Rohner and Smith (1996) enlarged broods of Great Horned Owls by 50 per cent, from two to three, and there was no apparent change in nestling growth rates apparently because parents supplied extra food. When they doubled the brood size to four, nestling growth rates fell below normal.

Another set of factors can mess up a statistical relationship between food supply and breeding success – 'density-*independent*' factors. A density-independent factor operates independently of predator densities. For example, there can be an increase in rabbits or voles, but the owls have poor breeding success anyway. Why is this?

Raptors cannot adjust to density-independent factors as easily as they can adjust to density-dependent factors. The best example of a density-independent factor is a rainstorm. Owls or falcons might arrange themselves over the habitat in pair densities relative to prey availability, a density-dependent factor, but rainstorms over two weeks stop *all* the owls from hunting and kill their just-fledged young independent of the density of raptors or food abundance. For example, Peregrine Falcons around Canberra produce *more* young during a drought, and *fewer* young during years of high rainfall. Frequent rain floods nests, kills young and stops adults from hunting. During a drought, the opposite is true. There is less chance of rain flooding a nest or decreased hunting time for adults. Consequently, many raptors produce more young in dry years than in wet years. Both density-*dependent* (food availability), and density-*independent* (rainfall killing young) factors determine the number of young they produce in a year.

Do the principles derived from Lack and applied to Northern Hemisphere owls, such as Snowy Owls, also apply to Australian owls? Australia is considered a 'boom or bust' continent, but most diurnal and nocturnal raptor species are strictly seasonal breeders. No study of any Australian owl has linked prey availability to breeding performance. Nocturnal prey species such as possums and gliders are assessable, and often they are counted in surveys, but none of these counts have, so far, shown a statistical relationship to larger or smaller clutch or brood sizes in Australian owls. There is scope for future study.

David Lack, the schoolteacher and eventual Oxford don, framed much of our thinking around stability and density dependence. Some of his notions have been questioned or altered, but many hold up. Later, he converted to Anglicanism and wrote a book called *Evolutionary Theory and Christian Belief* on the relationship between Christian faith and hard-line evolutionary theory, but this had little impact compared to his earlier works.

34

Tawny Owls in Kielder Forest

Kielder Forest on the Scottish border is a massive slab of conifers planted by the Forestry Commission in Northumberland between Hadrian's Wall and the Cheviot Hills. The lake at its centre is called Kielder Water. During the 13th to 16th centuries this region was the hangout of border reivers, warlike clans who profited from the struggle between England and Scotland. They stole livestock, food, anything not nailed down. Fortified farmhouses (bastles) and defensive tower houses (peels) dot the landscape.

Steve Petty studied Tawny Owls *Strix aluco* there. They are similar to Boobooks in that adults stay on one territory for several years. The breeding adults do not move around to follow irruptions of rodents as Snowy Owls or some Great Grey Owls do. They catch available prey on a long-held territory.

Strix means 'a kind of owl', especially a 'strident' one, a screech-owl, and *aluco* is derived from the Italian names *allocco* or *alocho* for Tawny Owl, originally from the Latin *ulucus*, a variant form of *ululus*, a screech-owl. This seems strange, but *Strix* was once the generic name for many owls including Barn Owls before taxonomists sorted them out.

Tawny Owls (Figure 34.1) range from North Africa and Europe to western Siberia. There is a strange empty gap before they turn up again in Turkestan, north India, east China and Taiwan. A pair was introduced near Napier, New Zealand in 1873, but the introduction failed.

These medium-sized 'wood owls' live in varied habitats, mixed forest with clearings, riverside forest, parks, larger gardens and farmlands with avenues of trees, as long as these habitats have enough trees to breed and roost in. Tawny Owls

Figure 34.1 Tawny Owl being measured (photo by Iñigo Zuberogoitia).

readily accept nest boxes, but some writers warn that nest boxes are not always a good idea. In most places Tawny Owls do not need boxes to prosper, and they are powerful predators of smaller owls, such as the Common Scops Owl, Eurasian Pygmy Owl, Little Owl and Tengmalm's Owls. These smaller owls can suffer when populations of Tawny Owls are elevated by the use of nest boxes. Scops Owls are now the rarest owls in Europe. Tawny Owls, some claim, are a factor in the extinction of the Eurasian Pygmy Owl in the Black Forest in Germany after World War II.

Male Tawny Owls in England of the race *sylvatica* weigh about 450 g, and females about 580 g, a little smaller than the Spotted Owls in Washington State (males about 570 g and females 650 g) but larger than Boobooks near Canberra (males about 280 g and females 320 g).

Studying Tawny Owls in Kielder Forest is different from studying Boobooks in Canberra. We brought light walking boots and our colleagues shook their heads – not the footwear for Kielder Forest. Researchers and farmers tramp through mud and standing water in Wellingtons. Kielder is cold, wet and sloppy and we slipped and fell on mud-slicked roads, on hillsides and between trees, as rain drifted over the study area.

Steve Petty based his study team at Deadwater Farm, a working property in hill country. The farmer and his wife served us breakfast and we watched Black Grouse courting on a steep hillside above the farm before we climbed into cars and drove to a series of wooden nest boxes nailed to trees. Steve accessed nest boxes by leaning a stepladder against the tree, a big difference from studying owls in Australia. Even without nest boxes, Tawny Owls found places to nest, in corvid nests, in raptor nests or on the ground at the base of a tree. Contrary to some claims, Steve thought that boxes in Kielder Forest did not significantly change the density of breeding Tawny Owls, it just made the nests and owls easier to study. Steve climbed up and brought down an incubating female to measure. He weighed this adult by looping a cord around its feet and hanging the owl from a spring balance, a technique we had not seen before.

Ian Newton picked up a dead Field Vole *Microtus agrestis* under the nest box, apparently killed then dropped by one of the owls. He said, 'This animal drives the whole system'. Ian studied zoology at Oxford University under David Lack and he applied many of Lack's principles to the study of raptors, after first applying them to British finches. As with David Lack, Ian was particularly interested in the ways that food limits animal populations, especially birds. In his 1998 book *Population Limitation in Birds* he proposed that pairs of birds could respond to increases or decreases in food supply in three ways:

1) The number of breeding pairs living in one area, and the number of young they produce, goes up with increasing food supply and down with decreasing food supply.
2) The number of breeding pairs living in one area stays the same, but the number of young they produce goes up with increasing food supply and down with decreasing food supply.
3) The number of breeding pairs living in one area and the number of young they produce stays the same each year. This happens when falcons, eagles, hawks or owls prey on many species, not one or two species, giving stability. They eat one prey when it is abundant, and then switch to another prey species when the first becomes scarce.

So, strategies 1 and 2 show a density-dependent relationship with annual changes in prey numbers. Strategy 3 shows no density-dependent relationship with annual changes in prey numbers, or at least none that is measurable. However, with strategy 3, the number of young might still increase or decrease from year to year because of density-independent factors such as rainfall.

The Tawny Owl is essentially an owl of forest habitat but is still successful in highly modified England, even where broad-leaved forest has been badly degraded. In Kielder Forest, patches of Sitka and Norway Spruce are cut every 40 to 60 years, leaving a mosaic of different aged stands of trees for owls to hunt and nest in – from mature trees, to no trees (clear-cut), to young trees, to middle-size trees and then to mature trees again ready for harvest. Tawny Owls in Kielder Forest live primarily on one species, the Field Vole, so it is easier to see density-dependent changes in Tawny Owl breeding success related to prey than, for example, with Boobooks and their many prey species in Canberra.

The Field Voles fed in the clear-cut areas until these clear-cuts grew back with trees, and then another forest area was clear-cut and this provided good vole habitat again. This appearance and disappearance of open areas for voles to feed on and breed in drove the vole populations up and down, so you might have a high vole population in one forest and a low vole population in a nearby forest in the same year.

Field Voles exhibited a three to four-year cycle of abundance. Steve measured breeding success or lack of success by comparing Tawny Owl clutch size to vole abundance. When vole numbers were low, Tawny Owls laid fewer eggs.

This fits one of Ian Newton's categories above, strategy 2, where pairs stay put and produce more or fewer young in relation to rising or falling food supply. They didn't move in low vole years to find new pockets of voles in other places, they rode out lows by reducing the number of eggs laid in relationship with a falling food supply dominated by a single species.

This relationship was not perfect. Density-dependent and density-independent factors are only two sets of factors driving the lives of owls, and they don't explain all breeding successes or failures. In the study area we first stopped at a Northern Goshawk *Accipiter gentilis* nest on a wooded slope. Goshawks nest in Canberra woodlands near Southern Boobooks, but these are Brown Goshawks and much smaller. The goshawks in Kielder were the most powerful avian predator in the forest, about the weight of an Australian Little Eagle. These Northern Goshawks were common in Britain during Roman and medieval times, but forests were felled and gamekeepers shot them because they killed pheasants, partridge and grouse meant for the landed gentry. Goshawks disappeared as a breeding species from the British Isles. When governments planted shady conifers across England, Scotland and Wales and new laws restrained gamekeepers, the goshawks returned, though not without help. Falconers shipped some from the European mainland and released them, sometimes with jesses on. Steve Petty watched goshawks increase during the 1980s in Kielder Forest and monitored them as they started killing Tawny Owls. Goshawks now were an important factor in breeding success.

A curtain of rain drifted over Kielder Water as we drove back to Deadwater Farm on a muddy track. We stopped at a wooden observation blind and watched, through slitted ports, ducks and other birds. At the far end of the lake, a single otter looping through shallows; glimpses of its curved back shone in sunlight.

35
Duelling and nest failures

Almost all examples of limiting factors in owl research describe *extrinsic* factors – food, enemies, parasites and nests. There is an assumption that animals are limited by declines in food supply or changes in other resources. That is, owls do not affect limiting factors, limiting factors affect owls. However, this assumes that all individuals in a population are the same. According to some researchers, individual differences in a population may lead to *self-regulation* via *intrinsic* changes in the population. The concept of *self-regulation* is mostly ignored in the raptor literature, sometimes maligned. Dennis Chitty argued that individual organisms can change over time so there will be a difference in mortality rates in a population in two different time periods – not because there are more predators or less food in one period than another, but because individuals have changed (Krebs 2001). Chitty (1960) argued that species are capable of regulating their own population densities without destroying renewable resources. The mechanisms that enable self-regulation involve mutual interference between individuals or some form of hostility between individuals of the same species, for example, spacing behaviour.

The main causes of nest failure in raptors are said to be lack of food, drenching rain, predators or chemical poisons. But there can be other reasons, such as interactions between close nesting pairs. Verreaux's Eagles nesting close together in Zimbabwe usually avoided adjacent nest sites (Newton 1979). When neighbours did use adjacent sites, their nests often failed. In addition, the disturbance caused by a new pair trying to break in and establish a new territory could destroy a breeding attempt. The pairs fight each other or take each other's food and this robs the pair of time and energy. Because of border conflict, some pairs don't breed and some abandon their eggs.

Through territoriality, raptors spread themselves out in relation to food supply and nest sites. Failures increase in the closest nests where there is too little food to support both pairs. Close nesting pairs fight or disturb each other in an effort to force their neighbour to move away and increase space between them – to spread out. This fighting is often linked to nest desertion.

Boobooks used territorial calling and Duelling to space themselves out across habitat in Canberra. In theory, density dependence could explain Duelling in Boobooks. Male Green (Chapter 23) forced his way into the territories of Males Red and White, there was antagonism and Red disappeared, leaving two pairs in the same space that once held three pairs. However, the *closest* nesting pairs in this study, Orange and White, had very *high* breeding success, while other pairs nesting further apart had *lower* breeding success. The important factor seemed to be Duelling, or conflict, not how close the nests were. How did owls interfere with each other and lower breeding success in neighbours?

The pairs Orange and White, with high breeding success, nested along a 'peaceful' border for many years. They Duelled along their other borders, 'hostile' borders with other owls, but never Duelled with each other. Sometimes other males in the study Duelled every night and after many nights of Duelling, one of the nests would fail. It looked as though males ruined each other's breeding attempts, killing the eggs or young of their adversaries or causing females to abandon them. Was there a clear relationship between Duelling and nest failure? To find out, we looked at data from 13 years and 1463 observation nights between January 1993 and December 2005.

We scored each Boobook nest as 'often Duelled 'when we saw Duelling on more than 10 per cent of nights, or 'seldom Duelled ' when we saw Duelling on fewer than 10 per cent of nights.

A 'nest failure' was defined as a mated pair copulating near or in a nest hollow, or laying eggs, then abandoning the breeding attempt. Pairs sometimes laid a second or third clutch when they failed, often in different hollows in the same territory. We counted each of these as a separate nest failure.

Results

The frequency of Duelling per observation night changed over the 13 years of the study (Figure 35.1). There were two peaks in Duelling: one around 1996–1997, another around 2004–2005. The two peaks corresponded with two increases in the percentage of nests that failed (Figure 35.1). The first peak related to years when Green placed his nest against the border of two other pairs and pushed further east each year until the male of one of the original pairs, Red, disappeared. The

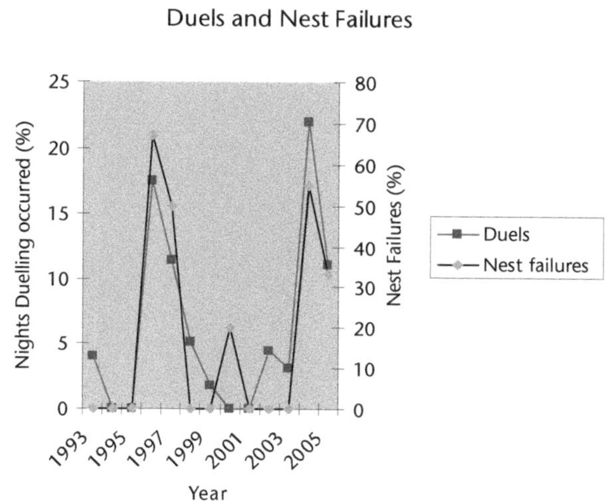

Figure 35.1 Per cent nights when Duelling observed per year and per cent nest failures per year August to December 1993–2005.

frequency of Duels flattened out between 1998 and 2003, and the number of nest failures stayed low (Figure 35.1). Duels and nest failures peaked again during 2004–2005 when the government pushed a new road through bushland territories of five pairs, destroying 8000 trees. Only two of five pairs remained after 2004.

Nests where males 'often Duelled' failed more frequently than nests where males 'seldom Duelled' (Table 35.1). Nests where males 'often Duelled' had a failure rate closer to 50 per cent, either because both males failed or because one succeeded and one failed. Changes in territorial borders often occurred near the sites of these Duels. You can see changes in nest site locations from 1993 to 2004 in Figures 35.2(a), (b) and (c).

Case Study 1 – Males Red, White, Green and Orange

We saw this pattern with a number of neighbouring pairs, but we can better understand it by looking at one case study.

As mentioned above, males Red and Green Duelled close to Green's nest from 1993 to 1997 as Green pushed further east into the centre of Red's territory, and

Table 35.1 Number of Successful versus Unsuccessful nests 1993–2005. n= 59 breeding attempts where males often Duelled (Duelling observed >10% nights) or seldom Duelled (Duelling observed <10% nights). Success was higher at nests where males seldom Duelled compared with nests where males often Duelled (c2 = 15.8, d.f. = 1, P<0.001)

Duelling	Unsuccessful	Successful
>10% nights	12	12
<10% nights	1	34

(a)

(b)

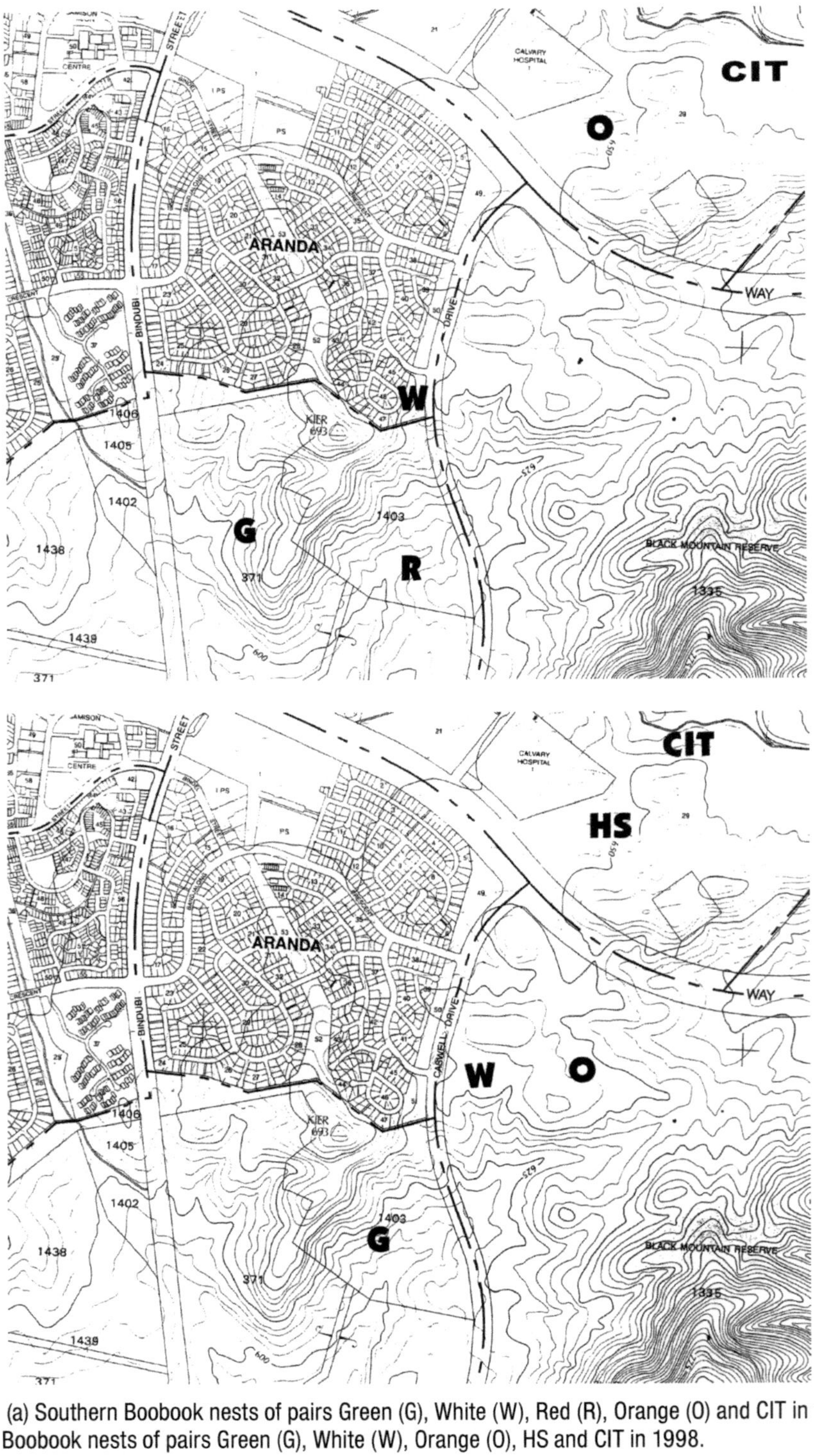

Figure 35.2 (a) Southern Boobook nests of pairs Green (G), White (W), Red (R), Orange (O) and CIT in 1993. (b) Southern Boobook nests of pairs Green (G), White (W), Orange (O), HS and CIT in 1998.

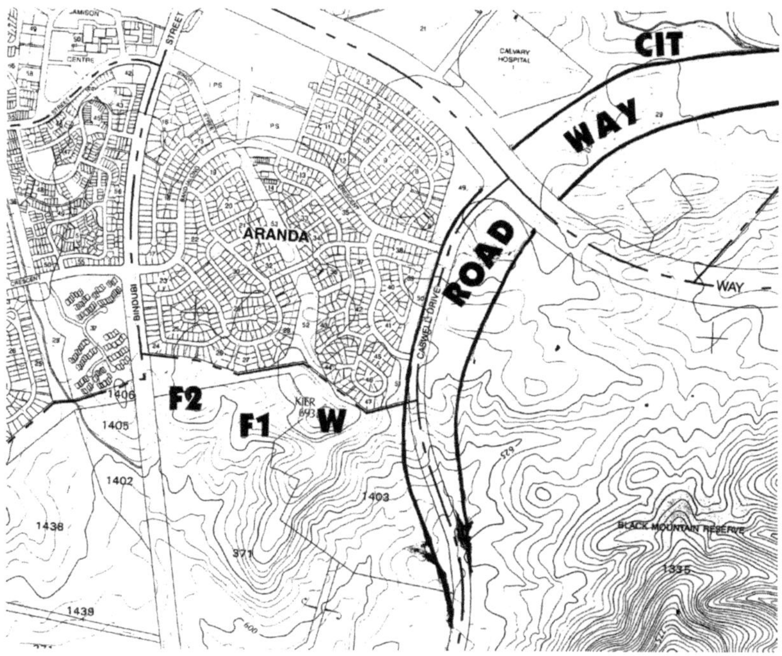

Figure 35.2 (continued) (c) Southern Boobook nests of pairs White (W), Floater (F1, F2) and CIT in 2004.

into a corner of White's territory (Figures 35.2(a) and (b)). In 1997, Green's original mate switched to White's territory and nested with him. Also in 1997, during the first peak in Duelling, Red persistently Duelled within 30 m of Green's nest, and apparently caused the Green pair to fail three times, once at the nestling stage and twice at the egg stage, adding to the number of failures we counted that year. White had a common border with Red from 1993 to 1998 and he Duelled with Green but never with Red. Red disappeared around 1 January 1998, and his mate was apparently killed by Green's mate on 1 September 1998 (Chapter 23).

When Red disappeared, Green and his mate absorbed Red's now abandoned territory. The Green pair fledged young there each year until 2004, the second peak in Duelling.

White fledged young in 12 out of 13 years from 1993 to 2005; he failed in 1996 during the first peak in Duelling, but fledged young during the second peak. By 2002 White and Orange, another neighbour, nested 200 m apart, but never Duelled and never had nest failures.

In 2004, after the government cut the new road through, Green and Orange disappeared. They may have been driven out, but we were not sure. White stayed but lost many foraging trees from the new road. To compensate for this, he absorbed the now empty territories of Orange and Green as shown in Figure 35.2(c).

In 2001, before the road was pushed through, a 'floater' appeared, a male with no mate or breeding territory, who was tolerated by White and Green, even though he occasionally called. In 2004, he attracted a female and they laid eggs in a hollow about 200 m from White's nest (see Figure 35.2(c)). The tolerance stopped.

White started long Duels every night on the hill just above Floater's nest and drove him and his new female out, causing nest failure. The new Floater pair tried again 300 m west (Figure 35.2(c)), away from White's nest. Duelling stopped.

White died on 13 February 2006, 13 years after he was banded. Maybe it was stress, maybe he was old. We guessed he was close to 20.

Case Study 2 – CIT

In 2004, after they felled trees for the new road, another long-breeding male, Hospital, disappeared (Figures 35.2(b) and (c)). His neighbour, CIT (Canberra Institute of Technology), filled the vacuum by doubling the size of his territory, just as White had tripled the size of his territory after the road was pushed through. This was apparently a density-dependent response to food shortage caused by loss of trees used for foraging. Three pairs along the roadway disappeared while two remained and bred on the new expanded territories.

As mentioned above, biologists attribute nest failures to increased fighting between closely nesting pairs, as we saw between Green and Red and between Green and White during the first peak in Duelling. During the second peak we saw it again between White and Floater, and between a number of other males, but this doesn't explain why Red nested close to White but only Duelled with Green. It doesn't explain why Orange and White never Duelled, and seldom had nest failures. Maybe these males were related.

After egg-laying we never saw female Boobooks involved in these disputes, only males, whereas disputes described for diurnal birds of prey, for example with Verreaux's Eagles, involved both sexes. The increase in Duelling and nest failure in 2004 and 2005 happened when pair densities were at their lowest, that is, when there was more room than ever before, but it probably was linked to food shortage caused by the destruction of trees.

In contrast to studies of diurnal raptors, none of the Boobook pairs seemed to suffer food shortage early in the breeding season. They all had enough food to initiate breeding and lay eggs, and Green had enough food to make three breeding attempts in 1997. The time and energy used in Duelling seemed to cause nest failure or disadvantage later in the breeding season, not early in the season. Females might leave these males. Territorial borders could change at sites containing important food sources used later on. That is, each male tried to maintain a territory border that enhanced his long-term success. For example, during the first peak in Duelling, White took his fledged young along with him to battle. His female had deserted and he was alone with his young. White and his juveniles camped out along Green's border, and White Duelled with Green at dusk. Later in the night White fed his young, or they fed themselves near this spot on the border. He balanced the costs of caring for his young against the costs of holding a food source that he might use in the following year.

What actually caused these nest failures? Some pairs seemed to be stressed by long bouts of Duelling, but how could this destroy a clutch of eggs? After all, it was just a calling contest. How could a male's voice destroy his neighbour's eggs or nestlings? Duelling looked harmless. They squared off, hooting and huffing at each other during the night along a border. Why didn't the male under pressure simply ignore his neighbour, leave the Duel, capture food for his female, come back and feed her? She was often food begging and females on eggs showed little interest in these boys' games. We could see no potential harm to the eggs or nest if a male simply broke off a Duel and left. The incubating or brooding female could not be cuckolded; it was too late for that. Females would kill a neighbouring male trying to enter her nest hollow. If a male left his mate alone, with the neighbouring male hooting on the border, so what?

We never saw males break off and leave in the middle of a Duel. Sometimes a male would stop calling, sit facing the challenger, as if watching and listening, but he never left to forage for his hungry mate. Our impression was that Duelling was a ritual with clear rules and consequences or penalties. If a male disengaged from a Duel, especially with a female looking on, it would cost him. The purpose of Duelling was to maintain territory borders and maybe to keep females. At least two females switched to neighbouring males after bouts of Duelling.

Maybe breaking off in the middle of a Duel represented a 'back down' with females looking on. Maybe a 'back down' increased the chances of a female leaving. Female raptors tend to switch territories more often than males do, and Boobooks in our study were no exception. Also, in our study there was more Duelling *after* the nest was occupied, not before it was occupied. There was also more Duelling *after* the female laid eggs. Maybe, with the females sitting on eggs, males could more easily destroy the breeding attempt of an enemy neighbour by engaging him in long Duels (these could last for hours) and stop enemy males from feeding their own females. Females might leave when their mates did not feed them, or they might 'plan ahead' and switch that year or next year to a stronger male who could win Duelling contests *and* keep her fed.

After females 'deserted' as normal, left their fledged young and travelled to a winter home range, giving males sole responsibility for the young, males still hammered away at each other in bouts of Duelling.

In summary, Duelling seemed to have three related functions: (1) to guard and maintain territory borders and food supply *over the long term*, (2) to destroy a neighbour's breeding attempt and (3) to attract and keep quality females.

Conservation

For many decades raptor biologists have tied the conservation of certain owl species to tall forests. The Powerful Owl is a symbol of the fight against exploitation of forests, especially against clear-felling. 'Save the forests and we can save the owl', or vice versa. To save the Powerful Owl, according to Stephen Garnett (1992), we need an accurate understanding of the densities of owl populations, how many pairs nest in a specified area of square kilometres and the lifetime habitat use by a pair. How long do they live? How much land does a pair, an individual or a family need?

Biologists in North American and Europe study owls in forests for the same reasons. The Spotted Owl in Washington State, the Great Grey Owl in Manitoba, the Eagle Owl in Europe, gone for a time from the United Kingdom but returned now, represent wild creatures that live in wild places, often in forest. The smaller *Athene* owls such as the Burrowing Owl and Little Owl present different conservation problems, and the *Athene* group is interesting because the genus contains only a few species and each presents different conservation problems. Case studies of all these owls can help us understand conservation.

36
Spotted Owls

Because of its status as a Flagship and Sentinel species, the Spotted Owl (Chapter 9) has been more thoroughly researched than most other owls. Still, its future is uncertain. Spotted Owls live in shaded forest and damp canyon bottoms in the North American west. Research on this species has influenced research protocols and conservation themes for owls in Australia and owls elsewhere in the world.

Washington State lies in the north-west corner of the continental United States just under Vancouver, British Columbia. Typical of western states it is big, 176 464 km^2, and it takes most of one day to cross by car. As you travel west from the mountainous Idaho border, the pine and fir forests grade into long flat stretches of blue-grey sagebrush and sandstone cliffs. Bald Eagles are common along rivers and Red-tailed Hawks and American Kestrels perch on telephone poles. Two-thirds of the way across you hit the forested wall of the Cascade Ranges running north-south, dividing the state. Roads through the Cascades can be icy in spring, covered in mist, with switchbacks, hairpin turns and rushing streams.

Loggers and conservationists fight over this country. These forests in western Washington State, Oregon and California are the strongholds of the Spotted Owl (see Figure 36.1). About the size of a bantam hen, Spotted Owls are chocolate and chestnut brown with irregular white spots over their backs and wings. As with most owls, the males are smaller, about 570 g, with females about 650 g, about twice the weight of a Southern Boobook and half the weight of a Powerful Owl.

Because it breeds and hunts in mature (commercially viable 2nd growth stands) and old-growth forest, the Spotted Owl has sparked a series of legal battles. In any town in the north-west you can hear both sides of the argument, and more than a

Figure 36.1 Spotted Owl being measured in Washington State.

few times you hear of a landowner threatening to shoot any Spotted Owl he discovers roosting on his or her land. This will stop greenies from alerting the government who would stop the landowner from cutting down trees, or so they claim. It is the right of an American landowner to log his own land, or it used to be his right until greenies took it away.

This fight, and the owls' association with old-growth forest, has made the Spotted Owl a model for wildlife conservation around the world. Biologists call such species 'Flagship Species' – animals that can be used as a charismatic poster child. Peregrine Falcons, Powerful Owls and Wedge-tailed Eagles are examples in Australia. Conservation organisations use eagles, falcons and owls as campaign logos, magazine covers and advertisements to attract funding and focus public attention on larger causes. Besides generating economic gains, Flagship Species are ideal tools for educating the public, their images capture public attention more effectively than do abstract concepts such as biodiversity or ecosystem integrity. The Spotted Owl is also a 'Sentinel Species' a reliable indicator of the condition of the environment because of its position at the top of the food web. Because such predators live at low densities, have low fecundity and have long periods of juvenile dependence, they are the first to disappear when habitat is degraded.

Of course, not all Americans buy this. The battle is felt most deeply along barricaded left-versus-redneck lines in the Pacific Northwest, and much of the battle has been waged in court. In 1973 the US Fish and Wildlife Service proposed that Spotted Owls be listed as a threatened species. It had been classified as threatened in Oregon since 1955. The first and second attempts to federally list the subspecies as Threatened or Endangered failed, so litigation started. In the 1988 Civil case No. C88-5737 – Spotted Owl and the Seattle Audubon Society versus Hodel, the Spotted Owl won; the court ruled that the decision not to list them was arbitrary and capricious. Some said the earlier postponement was driven by economics and the lack of quantitative data on population trends and habitat loss – there were no baseline data for the 1970s and 1980s. Plans were written to protect the Spotted Owl, then scrapped, then written again and mountains of money and thousands of hours given to research, much of it organised by the Spotted Owl expert Eric Forsman. More legal battles came, including a petition filed in 1993 to de-list the Northern Spotted Owl through all habitats in northern California. The petition failed.

There is another problem. Occasionally researchers find the closely related but larger and paler Barred Owl, a bird of forests in the eastern United States, in Spotted Owl habitat. It began creeping into western states and a biologist found it breeding there in 1974. Since then, researchers have found it breeding in places originally used by Spotted Owls. It has apparently pushed the Spotted Owls out, or used sites that Spotted Owls vacated and the two species have even hybridised, though this was a limited event. The hybrids look like larger, light-coloured Spotted Owls.

Lumber companies are now arguing that conservation efforts have failed, so they, the logging companies, should be allowed back into Spotted Owl habitat to cut trees. They, and owl researchers, are watching to see what happens next.

37

Athene and Burrowing Owls

The grey English winter had broken. Patches of green oak covered hilltops, their crowns like bunches of broccoli. Turned soil in fields the colour of chocolate would become hectares of yellow rapeseed by summer. Along a peaceful river you could smell chlorophyll rising from the grass, so different from Australia – no brown snakes, scent of eucalypt, yellow wattles or dead leaves crunching. It was strange to see a Southern Boobook there.

For many years, at Newent in Gloucestershire, Jemima Parry-Jones had run the National Birds of Prey Centre, and she had one of the biggest raptor collections in Europe. She successfully bred many of them and her father, Phillip Glasier, was England's favourite falconer. Jemima took over the Falconry Centre from her father in 1981. His book *As the Falcon Her Bells* led him onto movie sets to meet stars and English royalty and the book sparked a falconry obsession in quite a few American, British and Canadian children. The obsession nurtured falconers' clubs in North America and Europe, but the sport was suppressed in Australia, New Zealand, Norway and Sweden. They still sold falconry books in these countries, but the practice was illegal.

This hilly part of England had narrow lanes through grey stone villages and narrow alleyways. In slashing wind Jemima swung the lure in a wide circle, her skirt flapped around her legs. The Lanner, an African falcon, played with the lure, rose vertically above Jemima's head, flipped over using the wind as a tool and speared down to cut the lure out of the air.

The Eagle Owl *Bubo bubo*, meaning 'horned or hooting owl', a bigger cousin of the American Great Horned Owl, was next on show and the subject of a strange conservation story. Gamekeepers had shot Eagle Owls out of the United Kingdom, but now some were coming back, probably from Scandinavia, and starting to nest. The owl squatted and fought buffeting gusts with its weight, blinked in the sun and flew short distances, from a perch to Jemima's fist and back to a pole.

After the demonstration we walked through Jemima's raptor centre, and talked in front of the Boobook enclosure. We knew too much about Boobooks to feel comfortable about this one, alone in a cage a long way from home. Jemima said it was captive bred. She led us past incubators holding racks of eggs laid by hawks, eagles, falcons and owls and a massive steel meat grinder. She had a special interest in owls, wrote a book on them and showed us the egg of a Great Grey Owl, surprisingly small given the size of the owl.

We passed into her kitchen where a young Burrowing Owl ran loose. The tiny owl was bred at the Centre and Jemima had hatched it in an incubator, raised it and tamed it, using methods that would help the scampering owlet settle and reproduce in captivity.

The scientific name for the Burrowing Owl (Figure 37.1) is *Athene cunicularia*. The genus name *Athene* comes from Pallas Athena the daughter of Zeus in Greek mythology (Minerva in Roman mythology), the goddess of wisdom, the liberal arts and warfare. Traditionally Athena is depicted carrying a shield, an aegis and her favourite bird is perched on her shoulder – an owl in the ancient association with her primitive role as the goddess of night. The species name *cunicularia* comes from the Latin *cunicularius*, 'a miner or burrower'.

At the 2nd International Symposium on Owls held in Winnipeg in February 1997 a workshop on the vanishing Burrowing Owl on the Canadian prairies blamed its disappearance on degraded land, and on powerful new rodenticides on farms. Biologists had known for many years that Burrowing Owls were in trouble. Even A.C. Bent in his classic *Life Histories of North American Birds of Prey* said in 1938: 'It loves virgin prairies and unbroken plains but does not take kindly to cultivated land; consequently the encroachments of agriculture have greatly restricted its former range' (p. 384).

Burrowing Owls nest in burrows dug by squirrels, prairie dogs, badgers and other animals on the open plains, so the campaigns in the early 20th century to eradicate prairie dogs led to a rapid decline in Burrowing Owls.

Biologists had found 76 pairs in the province of Manitoba 15 years before, where now they found only one pair. Though biologists had reintroduced Burrowing Owls to British Columbia after 1979, only one breeding pair remained. The Canadian stronghold was southern Saskatchewan, but less than a thousand pairs remained there, down from 3000 pairs in 1978. And these were declining by about 10 per cent per year.

One speaker concluded that Burrowing Owl habitat was now too fragmented. These owls dashed around on bits of pasture and caught insect prey. Remaining patches of grassland were too small and fragmented. Burrowing Owls would not persist as a breeding species in Canada.

Speakers recommended a list of actions for saving the Burrowing Owl.

This owlet running loose in Jemima's kitchen would be sent to the United States for education purposes and to produce fledglings for release. It lived at Jemima's house with quite a few dogs, and ran through plastic pipe tunnels set along the floor, under the kitchen table and near the dogs' bed. As we sipped tea, it flew up on the table, ran across to Jemima for a piece of ground-up quail, then ran back to the wall, trilling the whole time. It ran back to her again, then to the end of the table, then back to her again for a cuddle, accepted a scratch under its ear with its eyes closed and its head tipped to one side. It fluttered down to the floor and teased one of the dogs, pecked at its wet nose, flew back up on the windowsill, head-winding at a fly crawling on the other side of the glass. It flew to a small desk, tucked up one of its long legs, closed its eyes and snoozed on the telephone.

Jemima said that her niece had left a loop bracelet coiled on the kitchen table and the owlet had gone berserk, leaping around and calling; the owl had even attacked it. Many nestling Burrowing Owls fall prey to rattlesnakes and this one seemed hard-wired to recognise snake-danger.

The genus *Athene* has only four members (six if you count the Lilith Owlet and Ethiopian Owlet as species instead of Little Owl subspecies, as do König and Weick 2008), and the group is an interesting case study in conservation issues – taxonomy, extinction, habitat destruction and local values. All four species look similar. When the taxonomic lumpers and splitters chop away at this genus they do little harm; sometimes they remove the Burrowing Owl, leaving three species instead of four, but the Burrowing Owl certainly looks like the other three and it remains there as I write.

Athene owls are small and have no ear tufts, not unlike little sandy-coloured Boobooks with longer legs. Facing you, they have wider more elliptical faces, sort of like Koalas, or distorted *Ninox* faces in a hall of curved mirrors at a county fair. Their wings are rounded like the wings of a Boobook. Eyes in all four species are yellow to sulphur-yellow.

Two of the Burrowing Owl's close relatives live in India. The Spotted Owlet *Athene brama* ranges from Iran to Vietnam, and south into the Indian subcontinent, the heart of its range. It lives in open or semi-open country, including semi-deserts, on the outskirts of villages and cultivated fields and in groves of old trees from lowlands up to 1400 m. It looks like a small, squat Burrowing Owl without the heavy spotting.

The second Indian species, the Forest Owlet *Athene blewitti*, is problematic. The scientific literature contains misinformation. In the book *Owls of the World*, published in 1973, edited by John Burton and written by a number of reputable scientists, a colour photograph of the 'Forest Owlet' is actually the common Spotted Owlet. It was first discovered in 1872 and was known from only four sites in deciduous woodland in western India. By the 1990s it had been absent from the scientific record for 113 years and some believed it was extinct, though other

Figure 37.1 Burrowing Owl pair in Idaho.

people claimed this was wrong. They had seen Forest Owlets more recently, or seen newer specimens in museums. A person named R. Meinertzhagen said his specimen of the Forest Owlet was taken some 220 km north of Bombay in 1914. But this specimen was faked. Pamela Rasmussen and Nigel Collar revealed how the Meinertzhagen specimen was actually taken in the 19th century, not in 1914. Somebody had restuffed the specimen, then relabelled it with false data. Rasmussen and Collar rediscovered the species in 1997, in a tiny region near Bombay in northern Maharashtra. The only Forest Owlets seen since then were near there.

The fourth *Athene* species, the Little Owl *Athene noctua*, is found across Eurasia and North Africa. In the following chapter we discuss how Little Owls were introduced to parts of the world outside their normal range. If biologists take the rare Forest Little Owl to England or America and breed it in captivity as a safeguard against losing the species, only the Spotted Owlet of the four *Athene* owls will be untravelled, the only *Athene* breeding essentially only where it evolved.

38
Little Owls

As mentioned above, the Little Owl *Athene noctua* brought the name *Athene* to the genus. Because of its early association with deities, including the goddess Athena, it became the sacred owl of the city-state of Athens. Little Owls were stamped onto the city's coins. The species name *noctua* comes from the Latin for the owl that was sacred to Minerva.

As with the other species in the genus, Little Owls are small and round, spotted or streaked on the breast and spotted on the back, with staring yellow eyes (see Figure 38.1). Their bold eyebrows create a fierce look, a frowning expression, as if disapproving of you watching them through binoculars.

Little Owls avoid cold or tropical countries. They range in a wide band stretching from southern England to northern Africa and east to China and Manchuria on the Pacific. Numbers are said to be decreasing in Switzerland, Belgium and parts of Germany, and their numbers drop at the end of severe winters. A very pale subspecies, a ghostly, sand-coloured form, inhabits the dry Biblical lands at the Eastern end of the Mediterranean. It may be the 'owl of the waste places' in the Bible, and the owl referred to in Psalm 102:6, 'I am like a pelican of the wilderness: I am like an owl of the desert'.

Little Owls know human-assisted travel, for example, to England. Ian Newton and Ian Wyllie from the Monks Wood Research Station near Abbots Ripton showed us a pair on farmland in Cambridgeshire. It was a gusty and rainy afternoon and Red Kites hung in the wind over fallow fields. This pair had a distinctive, undulating, bouncing sort of flight, like a woodpecker or an Australian Black-faced Cuckoo-Shrike.

Figure 38.1 Adult Little Owl being measured.

The pair of Little Owls perched with one foot up at midday in an ancient split oak. As we approached, one put its foot down and flew in the characteristic, bouncing, quick-winged flight, folding its wings tightly to its body between flaps. Across the bare field it pitched up into another oak. In Australia a dozen angry magpies, currawongs, Noisy Miners, and more would have attacked a Boobook, Powerful Owl or Barn Owl at midday. British birds ignored this owl.

Ian Wyllie said these owls caught earthworms on the tilled fields. The Little Owls we saw later in Spain, in dry farming country just outside Madrid, stood in the dirt outside their burrows in late afternoon with shadows slanting over them. Rabbits dug these nest burrows in a sandy bank, and the Little Owls stood with their heads bobbing, like tubby little Burrowing Owls. Iñigo Zuberogoitia, Inaki Castillo and Lander Astorkia showed us these owls nesting in abandoned buildings in the Basque country. The Spanish Little Owls took insects, chasing them over the dry ground.

Little Owls hunt more in daylight than do most owls, especially Australian owls. And they are one of only a handful of raptor species that people have introduced to other parts of the world. They were successfully introduced into England in the last quarter of the 19th century. At least some of these came from Germany.

These introductions took planning. The first attempt, by C. Waterton in 1842, failed. Waterton released five into Walton Park in Yorkshire but they all disappeared. Later, between 1874 and 1880, E.B. Meade-Waldo released 40 into Kent, with some success, but later releases in Norfolk and Sussex did not succeed. From 1888 to 1890 Lord Lilford introduced some from Holland into places near Oundle, in Northamptonshire. By 1891 the Little Owl's breeding range had spread to Rutland, and by 1892 into Bedfordshire. By 1923 birdwatchers could see Little Owls as far south as Lands End, and as far north as the Scottish border.

All this to introduce a feral species. Some, however, believed that Little Owls inhabited England as early as the Pleistocene, but were scraped off by glaciers and low temperatures in the Ice Age. Maybe the justification for introducing them to England went something like: 'Nature was cruel enough to remove them from England, so man can redress this mistake. If Little Owls occurred in England before the Ice Age, they belong here now'. But one wonders if the forest covering Great Britain after the Ice Age had excluded Little Owls as a species. The species, after all, needs open ground, and this open ground was artificial, laid down by farmers and woodcutters. As with most introductions, rabbits to Australia for example, people wanted them for amusement, or mistakenly thought they would control pests.

In England Little Owls often nest in willows and oaks, as the pair in Cambridgeshire did. They eat a few loved English mammals, such as hedgehogs and dormice, but the English forgive them. This was not always the case. During the 1930s, when the range of Little Owls had spread to its maximum, people started to complain. These introductions, they thought, were a mistake. As people do with any bird of prey, they accused Little Owls of killing game bird chicks, poultry and native songbirds. Loving the Little Owl as they did, some people loved poultry better, and Britain's gamekeepers wielded great influence. In the 1930s an owl rumoured to kill hand-raised or even wild-raised game preordained to be shot by the wealthy, should be exterminated. Gamekeepers shot all the Northern Goshawks, Sea-Eagles and Eagle Owls out of Britain, Scotland and Ireland, and some waged war against Hen Harriers and buzzards.

The accusation that Little Owls killed songbirds did not help. The British Trust for Ornithology sponsored an inquiry into Little Owl feeding habits and, after a 16-month study, concluded that Little Owls fed mainly on insects, adult and immature earwigs, craneflies, cockchafers and other beetles. The owls hunted partly in the day, especially at dusk, but chiefly at night, and they took small mammals up to the size of rats and small rabbits. They often perched on a post or tree limb watching for movement below. They caught some birds during the breeding season – Starlings, House Sparrows, Blackbirds, Song Thrushes and a few game bird chicks. The interesting point is that scientists were concerned about the Little Owl's impact on bird species that ornithologists and the public liked. They never asked the question – should a feral species like the Little Owl stay in Britain?

The next step in this man-assisted spread was New Zealand. This was successful, if successful is the appropriate word. As in England, the introduction took planning, and a lot of work. When scientists wrote up the Little Owl in the 1999 *Handbook of Australian, New Zealand, and Antarctic Birds* (HANZAB) nobody was certain about the race of Little Owls in New Zealand. They introduced some from Germany, the subspecies *vidalii*, but *vidalii* intergrades with the nominate subspecies *noctua* in southern Germany; so New Zealand Little Owls may be subspecies hybrids.

They brought Little Owls in to combat exotic passerines that raided orchards in central Otago. This failed. As we know from studies in England, they eat few birds, so did not control raiding finches. However, some argued that Little Owls did successfully control insect pests. In 1906, they brought in 28 more from Germany and released them – 14 at Ashley Downs and 14 at Alexandra – then in 1907 released another 39 in Alexandra. In 1908 they 'liberated' 80 more at various locations, and in 1910 imported 72 more, released 14, and sold the remaining 58 to farmers. They concentrated these releases on the South Island; the Little Owl was established in Otago by 1909. By the 1920s they had reached the southern tip of the South Island, and Stewart Island south of that. They were still increasing there in the 1950s.

Did they compete with Southern Boobooks in New Zealand? They lay their eggs around October (compared with April and May in England) in holes and crevices, trees, buildings, haystacks, cliffs, dirt banks and willow trees. They are said to have chicks around December and January (compared with June–July in England) about the same time as New Zealand Boobooks. They hunt at dusk and during the day, but mainly at night, so their hunting times overlap those of Boobooks.

HANZAB gives weights for Little Owls in New Zealand of about 167 g for adult males and 182 g for adult females. New Zealand Boobooks, called Moreporks, are about 177 g for males and 193 g for females, so they are close in size (see Table 18.1 in Chapter 18). One would think that introduced Little Owls would compete for territories with resident Boobooks, though Boobooks live more in forest and Little Owls in open country.

New Zealand ornithologists believe that Little Owls eat insects, spiders and earthworms, as in they do in tilled fields of Cambridgeshire, and probably some birds and mammals. In New Zealand you can witness entire scenes transplanted from England – a Little Owl resting in an English willow in a neatly trimmed English garden with an English hedge. Common Blackbirds and Starlings mob the Little Owl, the Blackbirds even knock it from its perch. It is strange how many English birds New Zealand Little Owls take – Skylark, Blackbird, Song Thrush, House Sparrow, Chaffinch, Greenfinch, Redpoll, Starling and more.

Should conservationists eradicate feral Little Owls from New Zealand and England? If conservation is about returning land to its native state – then yes. However, support for eradication would be poor. We are unlikely to remove feral horses from the Australian Alps, or Rainbow Trout from streams and rivers in Australia, or Pheasants and Red-legged Partridges from Britain and Canada, because people like having them around, even if they do not belong. An exception is the Masked Owl introduced to Lord Howe Island. Masked Owls probably wiped out the endemic Boobook there, and they now threaten rare bird species. Conservationists on Lord Howe plan to eradicate Masked Owls, but the other feral owl species around the world are probably safe. Introduced Little Owls will probably stay.

The genus *Athene*, then is an interesting case study in conservation – sorting out taxonomy, some facing extinction, habitat destruction and direct persecution. Should introduced forms be saved? König and Weick (2008) gave two Little Owl forms, the 'Lilith Owlet' on the Arabian Peninsula and the 'Ethiopian Little Owl' along the Red Sea, specific status. Are these forms subject to transport and re-introduction? Should different conservation policies apply to species versus races?

39

Conserving Australian owls

In a biological sense, *conservation* means to preserve a wide variety of species, subspecies and habitats (retain *biodiversity*) and to save the greatest number of each species appropriate to the ecological conditions in any given area (retain *densities*). If Australia loses more forest, prey species hunted by owls, open spaces or bushland, owl species may disappear. Biodiversity is reduced. If good habitat is taken up by houses, farms, mines or roads replacing food, shelter or nest sites, owls occur at lower breeding densities. This applies to subspecies. Of the seven Boobook subspecies in the region covered by Higgins (1999), two are extinct, leaving five subspecies (races).

Extinction

With mounting concern over global warming and habitat loss, there is considerable press given to the ultimate consequence of degrading habitat – extinction. What makes some species more secure, and others more at risk of extinction?

Extinction can be defined as when the distribution and abundance of a species or subspecies is reduced to zero. Newton (1998) said that the long-term security of an organism depends largely on three factors: (1) the extent of its geographical ***range***, (2) the number and extent of ***habitats*** the organism can occupy and (3) the average ***density*** of the species or subspecies within those habitats. Together, these factors determine the organism's overall numbers and security. Different combinations of these three variables give eight different classes of abundance (Rabinowitz 1981). Seven of these classes, 2–8, contain one or more elements of rarity. These elements are italicised and bolded below:

1) large range, wide habitat, high density
2) large range, wide habitat, ***low density***
3) large range, ***narrow habitat, low density***
4) large range, ***narrow habitat***, high density
5) ***small range***, wide habitat, high density
6) ***small range, narrow habitat***, high density
7) ***small range, narrow habitat, low density***
8) ***small range***, wide habitat, ***low density***.

On numbers alone, those species or subspecies in category (7) ***small range, narrow habitat, low density*** are most at risk. Consequently, three owls lost to us in the 20th century: the Lord Howe Island Boobook *Ninox novaeseelandiae albaria*, the Norfolk Island Boobook *Ninox novaeseelandiae undulata*, and the Laughing Owl *Sceloglaux albifacies*, were less secure than other owls. Island species and subspecies are more vulnerable than continental ones. More than 70 per cent of the bird extinctions recorded since 1600 AD were birds living on islands. Island species have been about 40 times more vulnerable to extinction from human causes in the last 400 years compared to continental species. Island endemics are more vulnerable because of their small geographical ranges (so they are more vulnerable to local destructive events, like hurricanes) and small population sizes (they can die out after losing relatively few individuals). Comparing different islands from around the world, the proportion of extinct island birds is inversely correlated to island size, with relatively more extinctions occurring on smaller islands than on larger islands, and more endemic species becoming extinct than non-endemic species (Newton 1998).

Given these eight categories, the endemic Christmas Island Hawk-Owl *N. natalis* in category (7) is our most potentially threatened owl. Its situation may become more desperate if the introduced Yellow Crazy Ant *Anoplolepis gracilipes* further threatens them. We could add to category (7) the Tasmanian Boobook *Ninox novaeseelandiae leucopsis*, especially if it is a distinct species *N. leucopsis*, the Tasmanian Masked Owl *Tyto novaehollandiae castanops*, and Tiwi Islands Masked Owl *Tyto novaehollandiae melvillensis*, because these are island forms and too little is known about their habits and abundance and they may be distinct species. Theoretically, these owls would be more at risk than continental owls such as Powerful Owls or mainland Masked Owls *Tyto novaehollandiae novaehollandiae*. The restricted-range subspecies of the Sooty Owl *Tyto tenebricosa multipunctata* in the Queensland Wet Tropics, and the 'Red Boobook' *Ninox novaeseelandiae lurida* in north Queensland, if they are a justifiable species or subspecies and genetic sequencing shows they are isolated from other forms, may also fit in category (7).

Some Australian owls fall into category (3) large range, *narrow habitat, low density.* The eastern Australian subspecies of the Rufous Owl *Ninox rufa queenslandica*, the southern subspecies of the Barking Owl *Ninox connivens connivens,* and all Australian subspecies of the Masked Owl may fall into this category, as well as the Powerful Owl and both subspecies (or species, depending on taxonomic treatment) of the Sooty Owl.

In 1994 the IUCN (International Union for the Conservation of Nature) devised 'Red List' categories for species at risk. To summarise the main categories:

Extinct – No reasonable doubt that the last individual has died (no verified record in the last 50 years).

Critically Endangered – Facing an extremely high risk of extinction in the wild in the immediate future.

Endangered – Not Critically Endangered but facing a high risk of extinction in the wild in the near future.

Vulnerable – Neither Critically Endangered nor Endangered but facing a high risk of extinction in the wild in the medium-term future.

Lower Risk – Does not satisfy the criteria for any of the above categories. Taxa in this category are separated into three subcategories:

1) **Conservation Dependent** – Cessation of continuing species-specific or habitat-specific conservation programs would result in the species qualifying for one of the threatened categories above within five years.
2) **Near Threatened** – Does not qualify for Conservation Dependent, but is close to qualifying for Vulnerable.
3) **Least Concern** – Does not qualify for Conservation Dependent or Near Threatened.

Data Deficient – Inadequate information to make a direct or indirect assessment of risk of extinction.

Using these criteria, Garnett *et al.* (2011) listed the Australian owls as:

Powerful Owl Least Concern
Rufous Owl (Cape York) Least Concern
Rufous Owl (eastern) Least Concern
Barking Owl (Torresian) Near Threatened
Barking Owl (southern) Near Threatened
Barking Owl (northern) Least Concern
Southern Boobook (Lord Howe Island) Extinct
Southern Boobook (Norfolk Island × New Zealand) Critically Endangered
Christmas Island Hawk-Owl Vulnerable
Sooty Owl Least Concern

Lesser Sooty Owl Least Concern
Masked Owl (Tasmanian) Endangered
Masked Owl (northern) Vulnerable
Masked Owl (Tiwi Islands) Endangered
Masked Owl (southern) Least Concern
Barn Owl Least Concern
Eastern Grass Owl Least Concern

Human threats to raptor populations

Declines in raptor populations can occur for at least four reasons: (1) contamination by toxic chemicals, (2) direct persecution, (3) predation and displacement by introduced species and (4) habitat loss.

1) Toxic chemicals

Owls in Australia have been less contaminated by toxic chemicals, such as DDT, than raptors elsewhere in the world, but the situation needs to be monitored. Second generation rodenticides, such as brodifacoum, difenacoum, bromadiolone (Bromakill) and flocoumafen were developed because first generation anti-coagulants, such as warfarin, required multiple doses to cause death as rodents developed resistance. These chemicals prevent blood clotting and cause death through internal bleeding. Second generation rodenticides are more toxic and persistent than first generation chemicals and they can kill owls (Newton *et al.* 2002; Debus 2009). Boobooks have died from brodifacoum campaigns against rodents in New Zealand (Stephenson *et al.* 1999). Other rodenticides include Pindone and 1080 used against rabbits. Aerial baiting with strychnine-laced grain to control mouse plagues is likely to cause secondary poisoning of Barn Owls and perhaps Boobooks and Grass Owls.

The rodenticide coumatetralyl, sold as Racumin, may be more owl-friendly. It is a multi-dose, first-generation anticoagulant, less toxic than brodifacoum, and has not developed problems of pest resistance to the same extent as the older rodenticide warfarin (Ratsak). Integrated pest control in the cane fields of northern Queensland, incorporating alternative strategies to the now-banned Klerat, and provision of nest boxes for owls, shows great promise for the control of rat damage and the restoration of the owl populations (Newton *et al.* 2002; Debus 2009).

2) Direct persecution

This includes shooting, trapping and intentional poisoning. Australian owls are less prone to direct persecution than are some diurnal species, like Peregrine Falcons that take domestic pigeons, or Wedge-tailed Eagles said to take lambs. Owls have a better public image.

3) Predation

Introduced predators probably killed off two of the three owls driven to extinction in our region. Mustelids and cats may have killed the ground-dwelling Laughing Owl in New Zealand. Introduced Tasmanian Masked Owls probably killed or displaced the Lord Howe Island Boobook (Higgins 1999; Garnett and Crowley 2000). The introduced Masked Owls should be removed from Lord Howe Island, and Southern Boobooks from a genetically similar population re-introduced there.

4) Habitat loss

Habitat is degraded in three ways.

1) The owl can be left with tiny fragments of habitat, for example, small patches of rainforest.
2) The owl can be left with a general alteration of its former habitat, for example, the farming of Australian grasslands. Both of these forms of degradation can affect biodiversity, but where fragments are left owls can be found within these fragments at no less a density than before. However, where general degradation occurs, the prey used by owls may be reduced. The owl species or subspecies still occurs over its original range but at greatly reduced densities.
3) Habitat can be totally destroyed and fragments are so small that owls can no longer live there.

Timber harvesting

All but two Australian owl species are woodland or forest species, so it is important to single out timber harvesting as a special threat. Kavanagh (2002b) noted that large forest owls occur at similar frequencies in unlogged and selectively logged areas, but they occur at lower frequencies in heavily logged areas. Where unlogged reserves were retained along streams and smaller drainage lines where Common Ringtail Possums occurred, the impact of logging on breeding owls was greatly reduced. Clearly there is more we need to understand about the effects of logging on owl populations, but all forest owls at the species level currently have Least Concern status, and some are successfully moving into major population centres, so the prognosis looks promising. However, some isolated woodland taxa, such as the Masked Owl in Tasmania and in the Kimberley are considered Endangered and Vulnerable respectively.

Wires, fences and power poles

Wires of any type constructed near nesting or hunting owls appear to increase the chance of injury or death. The dangers of wires, fences and power poles are twofold: (1) wires are difficult for a flying raptor to see and owls hit them and (2) powerlines present the extra danger of electrocution. Powerlines and fences have killed many owls. Barn Owls and Boobooks near Canberra have been found with their wings entangled in barbed fencing wires (Olsen 1992). In one US study, trained eagles were flown onto mock poles and wires. Slow motion films were used to determine exactly how raptors electrocuted themselves. It was estimated that 95 per cent of electrocutions could be avoided by changing only 2 per cent of the poles to a new design, altering the position of wires or adding a special perch above the wires for raptors to use. Dwyer and Mannan (2007) tested power poles near Harris Hawk *Parabuteo unicinctus* nests to determine which poles were lethal and, therefore, should be modified to prevent electrocutions. Juveniles were more likely to be electrocuted within 300 m of the nest, subadults and adults were electrocuted further away (201 to 300 m). Retrofitting poles within 300 m of the nest reduced the number of electrocutions.

40
Conservation techniques

Conservation techniques can be classed as **Preventive** or **Restorative**. Preventive techniques keep populations and habitats intact, before numbers are critically low. In this set of techniques, owls are retained in habitats close to those in which they evolved. Preventive techniques stop problems before the species or subspecies becomes seriously threatened, or handle problems when taxa show early signs of threat but have not seriously declined. Restorative techniques handle problems after a taxon has significantly declined and is at risk of being lost.

Preventive techniques

Reserves. As a general rule, owls require space to feel secure. They rarely cope well with close human interference and they rarely nest close to houses. Although large parks and botanic gardens can offer abundant prey, vegetation and security from human disturbance for roosting and hunting, they seldom offer enough space and security for breeding. In contrast, reserves can offer enough prey, vegetation and security for non-breeding and breeding owls. Kavanagh (2002b) noted that New South Wales and Victoria were the only states that had developed management plans on public forest lands (wood production forests and national parks) for large forest owls and strong legislation in both states protect these owls. For example, in East Gippsland, Victoria, 100 locations, each 500 to 800 ha in size, are designated special protection zones, primarily unlogged reserves for Powerful, Sooty and Masked Owls. Kavanagh (2002b) outlines conservation protocols used in New South Wales wood production forests; for example, exclusion buffers of 10 to 40 m around rainforest, wetlands, heath, rocky outcrops, etc. and the retention of hollow-bearing trees and dead trees.

Owls can also function as Umbrella, Bio-indicator or Sentinel species (see below) and indicate the best habitat patches to target for protection as reserves.

Research. All species of Australian owl need better study. Before a species or subspecies declines, we need an assessment of resources they need to survive. In general two types of research underpin sound, preventative techniques:

1) Research to uncover the causes of a species' or subspecies' vulnerability.
2) Research measuring which techniques most effectively conserve a species if it declines or becomes threatened.

Most owl conservation problems are linked to increasing human populations. The relationship between changes in habitat and human population growth are crucial. This may sound obvious, but some overseas species, for example Great Horned Owls *Bubo virginianus* in North America and Tawny Owls *Strix aluco* in England and Scotland, remain abundant in landscapes populated by humans. Are there Australian equivalents? We need research to determine which species and subspecies are harmed by increased human population, which are benefited, and why.

Bird-watchers and researchers

Disturbance by ornithologists and other researchers raises a vexed question. When ornithologists and authorities are told about a disturbance-prone nest, the location becomes widely known and more people visit the nest with the result that the pair is further disturbed. The best solution is public education (see below) and requests for care and cooperation, but precise nest locations of pairs at risk should remain secret. Conversely, if two groups of researchers are studying raptors in the same area, the particulars of nest locations should be shared with a view to minimising visits (Olsen 1992).

To minimise disruption to owls being studied, the following guidelines can be followed when visiting a nest site:

- have a clear idea of your data requirements; visit nest sites only as often as necessary to obtain these data.
- visit the nest with regard to weather, e.g. not during rain or on very cold nights when nestlings can be chilled.
- where possible, visit the nest in the latter half of the nestling period; make your visits short, 30 min or under unless prolonged observation is necessary. Take special care during the first week after hatching.
- individual birds vary; use common sense, be sensitive to their behaviour.

- where possible, approach the nest from the front, below and in clear view of the birds, talking as you approach so the owls hear you coming and are not surprised.
- flush the incubating or brooding owl with care so that eggs are not broken and young are not injured.
- for behavioural studies close to the nest: (i) be predictable, relaxed and avoid sudden movement or noises; (ii) gradually habituate birds to your presence; (iii) choose birds whose breeding success is known.
- do prolonged observation sessions after the owls are habituated to you and will forage and feed young in your presence. We find that certain female Boobooks refuse to hunt or feed in our presence, so we leave them alone.
- avoid disturbing the local habitat. Do not remove any vegetation near the nest.

Public education

Education may be our most powerful conservation tool. Owl conservation is enhanced when people learn about owls, see films about them, touch them, care for them and care about their habitat, and understand the dangers of disturbance, population growth, land clearance and pesticide use. Human interest stories strike a chord, particularly stories about altruism (caring for other people or animals), resolution of conflict, and human–animal interactions. Newspapers, radio stations and television channels willingly pick up stories about rehabilitated raptors, particularly orphaned young. And the media will publicise conflicts where a new road, housing development or industry may destroy raptor habitat.

Visiting a school with an injured owl can actively engage children in conservation. You can show students how to hold a raptor and, where possible, how to release it to the wild. Obviously no Australian owl can be released during midday when diurnal raptors, currawongs, magpies, kookaburras or other enemies can mob it. Teaching raptor rehabilitation to high school classes involves hands-on work and high levels of conservation awareness. Films such as John Young's *Wings of Silence* are excellent education tools. Booklets and pamphlets, games, kits and other conservation material help bird conservation, but the electronic media, glossy colour books and hands-on experience are the keys to engaging the public.

Legislation and protection

Legislation to protect owls in Australia is based on the premise that native birds are part of the public domain – they belong to the people. Sanctions imposed by protective legislation can be increased as part of an overall package, and we as citizens can influence this process through lobbying, the media or through local politicians. There continue to be infringements of raptor protection laws but fewer raptors are now shot or intentionally poisoned, and collectors take fewer eggs. The bigger problem is that farmers, developers and governments still break these laws. They destroy nests, eggs, nestlings, protective cover and prey used by legally protected owls. At the moment too little is done about these breaches of law.

Restorative techniques

Restorative techniques attempt to return owl populations that have decreased to critical levels. These can involve captive breeding, transporting species to healthier habitats or rebuilding habitats, for example with nest boxes, to increase the numbers of breeding pairs. Normally this means that owls are retained in dramatically altered environments from those the species evolved in, so it should never take precedent over saving good habitat (sometimes it does).

Research

Fundamental research on threatened and vulnerable species and subspecies underpins restorative techniques. The 'tools' include habitat maintenance, nests and shelter, food, captive breeding, rehabilitation and fostering.

1) Habitat maintenance

The most critical measure for conserving raptors is maintaining their residual habitat. Two measures are necessary: (i) find the remaining areas of good habitat and preserve them and (ii) increase the carrying capacity of these areas so they support more owls. These remnant areas should provide secure nest sites, shelter, food, space and freedom from disturbance.

2) Nests and shelter

In many places there is ample food for raptors but too few suitable nest sites and little shelter. Owls do not build their own nests; they use ledges in caves, buildings or tree hollows. They will often accept a man-made structure that resembles their natural nest. Nest boxes are used throughout Europe where forests have long disappeared. Certain owls like Boobooks, Masked Owls and Barn Owls in over-cleared areas lacking suitable hollows may use them. Nest boxes can play an important role in species and population conservation as long as the chosen sites are protected from inclement weather, direct sunlight, predators and human interference. However, nest boxes are no substitute for maintaining natural nest sites in natural habitat; this usually means retaining natural tree hollows and woodland. While nest boxes have become the norm in places like Europe, used in place of natural habitat, in Australia, nest boxes and habitat restoration should only be used as a last resort. They should never be used as a political trade-off for the destruction of better habitat elsewhere. This sometimes happens. Ed McNabb and his colleagues in Victoria have tested various designs and fledged young from boxes in Melbourne suburbs lacking natural hollows (McNabb and Greenwood 2011).

3) Food

Availability of food and nest sites sets the carrying capacity of land containing owls (Newton 1979). Usually it is harder to raise the carrying capacity of a piece of land by increasing food supply than by increasing the number of available nest sites. Increasing food supply usually entails changing land use to promote an increase in prey numbers. Again, this is best achieved by preserving good habitat in the first place – on farms, in cities and in nature parks and other reserves.

It is crucial that biologists assess habitat quality for prey, including introduced species such as European Rabbits, Common Starlings, Common Mynas, Black Rats, House Mice and House Sparrows that can benefit some Australian owls. Fortunately, these introduced prey have been less contaminated by organochlorine pesticides here than in other parts of the world. Paradoxically, abundant introduced prey species can lift owl densities in modified environments, on farms or near cities, to higher levels than in 'natural' environments that lack these introduced prey species.

Habitat restoration and corridors

Debus (2009) noted that restoration of some indigenous tree cover on formerly wooded rural land is a necessary conservation technique because the large owls may be reluctant to traverse extensive open country. He suggests that the future of the Barking Owl in the sheep–wheat belt may rely on the establishment of corridors linking remnant patches of habitat and on the re-establishment of patches of native woodland. Corridors planned by some city governments are too small to accommodate large breeding raptors.

Captive breeding, rehabilitation and fostering

In general, owls reproduce at lower rates than other birds. Many owls take several years to reach reproductive maturity and most species breed only once per year at best, some *Tyto* species excluded. Clutch sizes in *Ninox* are relatively small. Using certain techniques that increase the number of young fledged, or the number of adults in a wild population, can offset this naturally low productivity.

David Fleay pioneered the captive breeding of Australian owls, but the current Australian expert is Rick Webb at Featherdale Wildlife Park. He and his colleagues have bred most or all of the Australian species.

At the moment, no Australian owl populations would be noticeably increased by rehabilitation or captive breeding. Crippled *Tyto* owls breed well in captivity, and a carefully planned and executed captive-breeding and release program could help recover decimated wild populations if these owls do decline, as was done with Barn Owls *Tyto alba* in England. Debus (2009) argued that we should, perhaps, retain captive stock of the Christmas Island Hawk-Owl for release in a new location if Crazy Ants affect their numbers. Some crippled *Ninox* owls breed readily in captivity, but the hacking method of releasing captive-bred *Ninox* needs to be refined. Some *Ninox* juveniles follow their parents around inside the territory and don't receive prey deliveries at a central nest (Olsen 1990), so they may not return to a single hack box after fledging (see Appendix 2).

Double clutching

Removing a clutch of eggs from a pair of wild or captive owls induces some owls to lay again. The first clutch is hatched in an incubator, the second is left with the wild or captive owls. Effectively, the pair produces twice as many young as they normally would. In a variant of this technique a captive pair of owls that hasn't laid eggs is given a 'dummy' clutch to start the incubation process, then given a normal clutch from another captive pair. The other pair lays a new clutch and both pairs hatch the eggs and raise the young as their own.

Fostering

Fostering is the placing of young owls hatched in captivity, or in the wild, into the nest of another pair of owls with no young, or with a small brood. Wild diurnal raptors, like Peregrine Falcons, readily accept fostered young, particularly if they already have nestlings about the same age. However, Olsen *et al.* (2002a) suggested that *Ninox* owls, unlike most diurnal raptors, can recognise the calls of their own young. This could work against successful fostering. McNabb (1996) fostered two Powerful Owl fledglings to adult pairs in the wild, so, under certain circumstances, *Ninox* owls do accept nestlings and fledglings other than their own. Fostering and hack techniques with *Ninox* may differ slightly from those used with other raptors, and may need refining.

Using owls to conserve non-owl species

Birds of prey are used as 'surrogate species' to help conserve other plants and animals. Conservationists save habitat needed by eagles, hawks, falcons or owls in order to save other species of bird, tree, grass, reptile, insect and so on. The argument is that birds of prey as top predators require space, and through predation, they affect other animals and plants at lower trophic levels, so conservationists can save many plant and animal species at lower trophic levels by saving raptor species. Furthermore, some raptors look 'glamorous' compared with other animals, so they can be used as conservation tools to draw attention to vulnerable species and areas, focus government or public attention on areas or taxa that would otherwise be difficult to conserve. Hence, the raptor is a 'surrogate', a 'substitute' for large, costly studies of biodiversity. Saving top predators saves species richness.

Raptors can be used as 'surrogates' in at least five ways, as: (1) Biodiversity Indicators, (2) Keystone Species, (3) Umbrella Species, (4) Sentinel Species or (5) Flagship Species (Caro and O'Doherty 1999).

1) Biodiversity Indicators

Areas with breeding raptors are said to contain more biodiversity (species richness) than same-sized areas without breeding raptors. This concept is central to claims where raptors are used as 'surrogate' species, so we will look at one example in detail.

Sergio *et al.* (2006) used six raptor species to determine if areas where raptors nested had higher species richness than similar-sized areas where raptors did not nest.

1) They first located the nests of Northern Goshawks *Accipiter gentilis*, Tengmalm's Owls *Aegolius funereus*, Eurasian Pygmy Owls *Glaucidium passerinum*, Tawny Owls, Long-eared Owls *Asio otus* and Scops Owls in the Italian Alps.
2) For control sites they chose: (i) sites randomly selected in comparable habitat (spatial controls), (ii) sites containing nests of randomly selected bird species at lower trophic level (taxonomic A controls) and (iii) sites containing six bird species, other than the raptors, with specialised ecological requirements (taxonomic B controls).
3) For biodiversity measures (dependent variables) they compared the biodiversity (species richness) at sites occupied by breeding raptors with the biodiversity at control sites. They walked transects and did point counts at sites with raptor nests, and at control sites to determine: (i) diversity of bird species, (ii) diversity of butterflies (Rhopalocera) and (iii) number of tree species at each point count.

Results – When they compared the spatial, taxonomic A and taxonomic B controls with the sites that contained nesting raptors, they found that sites with nesting raptors contained: (i) more bird species, (ii) more vulnerable bird species, (iii) more tree species (except 1 site) and (iv) higher bird diversity (species richness, measured using a statistical technique called the Shannon Index). Also, grassland and farmland with breeding Scops Owls and Long-eared Owls had more butterfly species, and sites holding the six breeding raptor species had higher *densities* of birds and butterflies.

To people who know about owls, the use of bird species, butterflies and trees as dependent variables must seem a little odd. Why count the number of bird species like warblers, robins or woodpeckers in an area containing Tawny Owls and

Long-eared Owls? These are mammal specialists; they live mainly on small mammals such as voles, not birds. Researchers do this because species richness is often *indirectly* related to the raptor, not directly related to it; the researchers look for standard *indicators of species richness*, and these include birds, butterflies and trees, even in home ranges of owls that consume only mammals.

It would be unfair to say that the Sergio *et al.* (2006) study went unchallenged. The study was attacked by at least three groups of researchers who believed that home ranges of top predators, such as owls, did not necessarily show more species richness (biodiversity) or higher bird densities than sites selected from bird atlas surveys. They argued that bird atlases (surveys) were a better way to find species richness, species 'hotspots'. The debate continues, but more evidence is accumulating that saving areas used by breeding raptors does save other species (Sergio *et al.* 2008).

2) *Keystone Species*

The raptor has disproportionate impact on the environment relative to its abundance because of its predatory habits. For example, Eagle Owls in the European Alps might depress the number of Tawny Owls in an area, and high numbers of Tawny Owls depress the number of Scops Owls, so keeping Eagle Owl numbers high keeps Tawny Owl numbers low, and this keeps Scops Owl numbers high. A problem with the concept of 'Keystone' species is that little evidence in the literature supports it. It applies to certain circumstances but not others, and it can be difficult to predict accurately how a predator will affect species at lower trophic levels in its home range. Ecosystems are complex.

3) *Umbrella Species*

Raptors need large areas to breed and hunt in, so saving raptor home ranges also saves animals and plants inside these large home ranges – it saves the smaller home ranges used by smaller species. In contrast to using raptors as Biodiversity Indicators and Keystone Species (above) where researchers select a certain area as a good place for a reserve because it contains breeding raptors, and raptors affect biodiversity through predation, this measure is about *size*. Umbrella Species provide large areas with more biodiversity (species richness) not so much because they have more species than other areas of the same size, but simply because they are large. For example, higher biodiversity was associated with expanses of habitat classified as suitable for Eagle Owls in the Alps (Sergio *et al.* 2004) because Eagle Owls are large predators that need big territories to hunt and breed in.

There has been little quantitative investigation of the effectiveness of top predators as Umbrella Species but the umbrella role may be useful to help us 'think large', thus promoting larger reserve size (Ray *et al.* 2005).

4) Sentinel Species

Raptors are sensitive to human-induced changes in environment, so studies of breeding raptors can detect conditions in the environment before these changes affect other species. Raptors are said to be reliable indicators of the condition of the environment because of their position at the top of the food web, and because they live at low densities, have low fecundity, have long periods of juvenile dependence and are the first to disappear when habitat is degraded. This makes them more vulnerable to human-induced changes in ecosystems than some other animals. The best-known example is the research on toxic pollutants like DDT and PCB that triggered extensive investigations in the 1950s and 1960s because of their effects on bird and fish-eating raptors. The British scientist, Derek Ratcliffe found eggshell thinning in Peregrine Falcons, and his research was used to stop pesticide use in North America, Europe and Australia. These investigations eventually showed wide-ranging effects on other birds, and these products were subsequently banned (Ratcliffe 1993). More recently, owls have signalled problems with rodenticides in the environment (as discussed in Chapter 39).

5) Flagship Species

Raptors, including owls, can be charismatic poster species. Conservation organisations use owls as logos (for example, the Australasian Raptor Association), magazine covers and advertisements to attract funding and focus public attention on larger causes. In the 1960s the Peregrine Falcon was not only a Sentinel Species (see above) but also an icon of conservation campaigns to ban pesticides. The Spotted Owl, also a bio-indicator (Sentinel Species), was used as a Flagship Species in campaigns in the western United States to save old-growth forest (Newton 1979; Newton *et al.* 2002).

Besides economic gains, Flagship Species are ideal tools for public education. Charismatic species capture public attention more effectively than do abstract concepts such as biodiversity or ecosystem integrity.

Owls work as Flagship Species. However, some might argue that conservation already concentrates too much time and money on charismatic species, and conservation needs to be driven, instead, by ecological need. Some animals attract millions of dollars of public and government funding simply because they are charismatic, not because they are rare. The argument is that conservation should be directed by research findings, not by public fashion. In addition, Australia lost two Boobook subspecies in Australia at the same time that considerable money and effort was being spent researching Wedge-tailed Eagles, Powerful Owls and Peregrine Falcons, charismatic species that were not threatened. Hopefully we will not make this mistake a second time. Wedge-tailed Eagles, Powerful Owls and Peregrine Falcons are not endangered in Australia, but we have almost no information on the status of island owl species and subspecies in Australia and the region.

In summary, owls are effective surrogate species, They can lever public support, raise funds, preserve biodiversity, protect and prioritise reserve sites, and help in the planning, size and configuration of protected areas (Gittleman *et al.* 2001; Ray 2005). But we need further research to validate the value of owls as Umbrella Species, Keystone Species, and Biodiversity Indicators and the potentially adverse affects of concentrating on secure continental species to the detriment of island forms.

Conclusions

The future for owls on mainland Australia and in Tasmania looks bright. The research efforts of Rod Kavanagh, Rohan Bilney, Chris Pavey, Stephen Debus, Ed McNabb, Nick Mooney, Fiona Hogan, Raylene Cooke, Todd Soderquist, Mathew Stanton and others are giving us the knowledge we need to conserve large owls in wood-production forests, remnant woodland, reserves and on farmland. Repeat surveys continue to find forest owl species across logging mosaics, showing these owls can survive and breed when some habitat is altered.

Australian biologists are preparing and implementing recovery plans for large forest and woodland owls in the south-eastern states, and for Grass Owls in New South Wales. Such plans are still needed for threatened owls in Queensland, South Australia and Western Australia. A recovery plan is also in place for the Christmas Island Hawk-Owl and is being updated (Debus (2009).

Debus (2009) argues that a Landcare ethic should become universal in rural Australia. Otherwise, we will lose ecosystems that provide our food, fibre and water as well as our precious owl biodiversity. Owls can be used as Umbrella, Sentinel and Flagship Species as well as Biodiversity Indicators to help save this land. They are at the forefront of our fight to save tropical forests and woodlands in northern Australia, the next important conservation arena.

Wallacea

Wallacea lies north of Australia, close enough for Australian biologists to access. It is a crucible of biodiversity. Bird species there amplify and sharpen conservation issues – taxonomy, extinction, habitat destruction and local values.

41
Wallacea

For over a decade the island of Sumba in Wallacea hid a mystery. Only a few ornithologists had glimpsed the 'mystery owl', and they disagreed about it. Some claimed it was a hawk-owl in the genus *Ninox* like the Southern Boobook; some claimed it was a Scops owl in the genus *Otus*, like the Common Scops Owl in Europe.

Looking at a map of eastern Indonesia (Figure 41.1), it is not clear why qualified ornithologists believed a mystery owl on Sumba was closely related to owls in far away Europe or Asia, not to owls in Australia. Australia is closer to these islands than it is to New Zealand, Fiji, New Caledonia or Solomon Islands. However, Australia, New Zealand, Fiji, New Caledonia and Solomon Islands are part of the *Australasian avifaunal region* where mainly Australasian birds evolved and live. In contrast, Bali and Java lie in the *Oriental avifauna region* close to Sumba. The chain of islands stretching east from Java to the island of Timor form a necklace joined to South-East Asia. The islands of Timor, Sumba and Flores are close to Australia but linked by stepping-stones to Asia, not to Australia. Sumba lies within the Australasian avifaunal region, but many birds there are close to Eurasian birds – and can even be the same species.

Birds can fly from island to island, so you might expect Australian-origin birds on Sumba, Flores and Timor, and some Asian-origin birds because of links to Asia. This is the case. For example, there are no woodpeckers in Australia, but many woodpecker species in Europe, Asia and the Americas, and several woodpecker species on Sumatra and Java. The genus *Dendrocopus*, woodpeckers you find in England, occurs on Flores, close to Sumba. You can stand on the east coast of Sumba with binoculars and see Flores. Bird-watchers from England are familiar

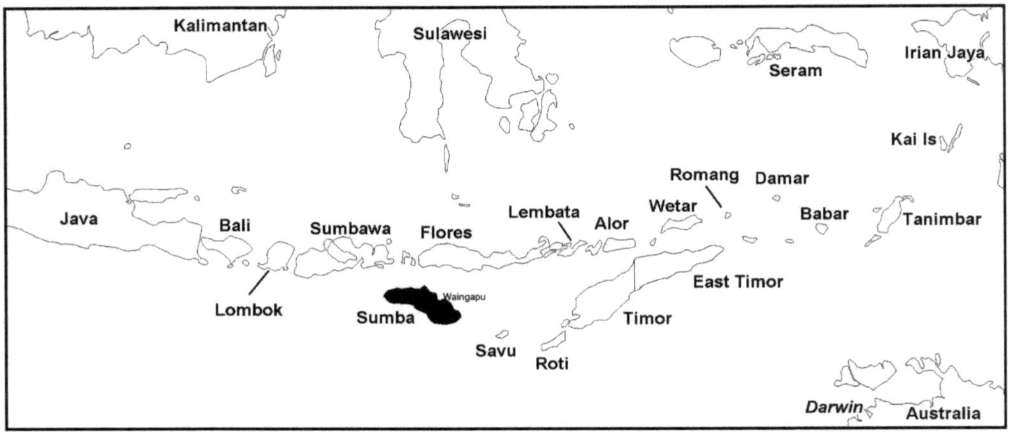

Figure 41.1 Islands of Sumba and eastern Indonesia.

with a number of bird species on Flores because they know them or their close relatives from London, Cambridge or Scotland. Yellow-crested Cockatoos, a smaller version of the Sulphur-crested Cockatoo of Australia, also live on Flores. Bird-watchers from England or Australia can identify many birds on Sumba and Flores.

Common English birds, such as the Great Tit, breed and flit through oak forest on Sumba. In England, Eurasian Sparrowhawks chase Great Tits as prey. On Sumba, Australasian Brown Goshawks chase Great Tits as prey.

Grey Herons, a species tourists see in central London, stalk through swamps on Sumba. Nesting close to them you might see White-faced Herons, an Australian species. On Sumba the Sacred Kingfisher is no surprise to Australian bird-watchers, but Common Kingfishers, the small, jewel-like divers known to British bird-watchers, also breed on Sumba.

This region of mixing, *Wallacea*, is bounded by the Philippines to the north, New Guinea to the east, Australia to the south, and Bali to the west. In Wallacea, the Eurasian avifauna crushing down from the north presses against the Australasian avifauna crowding up from the south. Why does this region of mixing start east of Bali, and why don't scientists include western Indonesian – Bali, Java and Sumatra – in the region called Wallacea?

We need to understand something about Alfred Russel Wallace, for whom Wallacea is named. He was a bit of a misfit, awkward and shy, maybe a little uncomfortable living in Victorian England. From 20 April 1854 to 1 April 1862 he travelled by boat to islands in the Malay Archipelago, today's Indonesia, collecting bird and butterfly specimens for private collections and to supply specimens to museums. In 1858 Wallace sent a letter to his colleague Charles Darwin containing a paper Wallace had written called 'On the Tendency of Varieties to Depart

Indefinitely from the Original Type'. The paper outlined in detail the origin of species by natural selection, the tendency of species to diverge from a common ancestor – the idea that Darwin had taken his adult life to develop (van Oosterzee 1997). It had come in a sudden flash of insight to Wallace during a bout of malaria on these remote islands, and it must have shocked Darwin because he had never published his theory.

The Linnean Society meeting of July 1858 announced the joint 'Theory of Evolution' by Charles Darwin and Alfred Russel Wallace. Wallace urged Darwin to publish his manuscript *Origin of Species*, which he did in November 1859.

When he visited the island of Bali in 1858, Wallace saw much that was familiar. Bali shares 97 per cent of its bird species with neighbouring Java and he knew the birds of Java. Eurasian species like barbets, pheasants and woodpeckers occur there.

When Wallace crossed the narrow channel from Bali east to Lombok, he was astounded. 'On crossing over to Lombok … I … found a totally different set of species, most of which were utterly unknown not only on Java, but also on Borneo, Sumatra, and Malacca. For example, among the commonest birds in Lombok were white cockatoos and three species of honeysuckers, belonging to family groups entirely absent to the west.'

Actually, Lombok shares 50 per cent of its bird species with Bali, but the other 50 per cent of its birds differ from Bali, a marked contrast to the 97 per cent shared by Java and Bali.

During the Pleistocene Ice Age, water was sucked into the polar ice caps causing sea levels to fall. The dropping sea levels linked adjacent islands, like Java and Sumatra, making a single land bridge to Asia. Falling sea levels may have linked islands such as Lombok, Sumbawa and Flores to each other but kept them split from Asia. A deep ocean trench runs between Bali and Lombok so Bali and Lombok were never joined, and mammals in particular were blocked from travelling between the two islands. Marsupials were blocked from travelling north from Australia, and tigers, leopards and squirrels blocked from travelling south into Australia.

The Wallace Line cuts between the close islands of Bali and Lombok. West of Wallace's Line you are more likely to see tigers, leopards, elephants, squirrels, martens, wolves or rhinoceros. East of the line you are more likely to see kangaroos or possums.

Birds can more easily cross from island to island, but west of Wallace's Line, or at least close to it, you are more likely to see Eurasian avifauna, like giant eagle owls in the genus *Bubo*, as well as woodpeckers. East of Wallace's Line you see more birds like Australian lorikeets or hawk-owls, *Ninox*. Wallacea is the area where the two bird groups collide and mix, and this makes islands like Sumba fascinating to ornithologists.

On Sumba you don't see Welcome Swallows, the common, blue-backed Australian swallow with a long, forked tail, but you do see Barn Swallows, the common swallow from Europe, England and North America. Most of these spend the northern winter on Sumba, then leave.

Three other characteristics of Wallacea make the islands important to ornithologists. First, the islands carry many endemic species, birds found only on one island and nowhere else in the world. Island endemics are important to the study of evolution, and crucial to species conservation because they are at greater risk of extinction than species distributed across great continents.

Second, there is little known about species richness in Wallacea – ornithologically, it is unexplored. If you collapse the islands of Wallacea, take away the water that separates them, it is a relatively tiny landmass, much smaller than Australia. But, so far, ornithologists have described about 700 species of birds, close to the number described for Australia. In time, ornithologists will describe more new species from Wallacea, and it is impossible to predict whether these will be taxonomically close to birds in Australia, Europe or neither. Some now described as subspecies will turn out to be full species.

Third, little is known about the biology of species already described for Wallacea. Ornithologists have yet to describe the nest and breeding habits of many species known only from one or two skins in a dusty drawer in Europe, especially the owls. There is also much to learn about migratory patterns of Australian birds in Wallacea. The Black-faced Cuckoo-shrike, a common bird in Canberra during summer, disappears in winter, as do some of the bird-catching Hobbies. Many fly north to Wallacea during the austral winter. Red-capped Plovers are Australian waders that appear to migrate north, and experts said they wintered but did not breed in Wallacea. But we found them breeding on salt lakes near Dili, East Timor, the first breeding records for Wallacea. It is relatively easy to make new discoveries in this unexplored region.

Australian raptors such as Spotted Harriers and Australian Hobbies breed on Sumba. You might expect common Australian raptors such as Brown Falcons or Wedge-tailed Eagles on Timor or Sumba, because all three live in New Guinea on the edge of Wallacea. They do not. Instead, an isolated population of powerful Bonelli's Eagles lives there, a bird common around the Mediterranean, North Africa and Asia but unknown in Australia. Another eagle, the snake-eating Short-toed Eagle from Eurasia, is unknown as a breeder in western Indonesia and completely unknown in Australia, but it nests on Timor and Sumba.

Why do these two eagles have isolated populations on Timor and Sumba, far away from the nearest Bonelli's and Short-toed Eagles in India and Asia? Nobody has given a convincing explanation. Many bird species in Wallacea should not be there, or we cannot logically see why they live there.

The rules are not clear.

42

Sumba Barn Owls

The potato-shaped island of Sumba lies at the crosshairs of latitude 10°S and longitude 120°E in Wallacea. It is south of the islands of Flores and Sumbawa in the Lesser Sunda chain of Indonesia. Sumba is about 210 km long with a surface area of 10 854 km^2.

At 1225 m, the highest point is Gunung (meaning Mount) Wanggameti. The island experiences a dry winter season from May to November and a wet summer season around December to April. Annual rainfall is between 500 mm on the south coast and 2000 mm on the inland hills. Closed-canopy forest, comprising mostly deciduous trees but also some evergreen, covers less than 11 per cent of the island and is confined to small pockets because locals repeatedly clear and burn vegetation for grazing and cultivation. These deciduous trees drop their leaves during the dry season, as northern oaks do in winter; skeletal, leafless trees in the forest that look dead.

A central plateau on Sumba formed of limestone and sandstone is cut in places by deep gorges. At night while looking for owls, it was cooler up there than down along the coast. Sumba is dotted with freshwater lakes, dependent on rainfall in the wet season. The lakes form in scoops of land on the high plateau and in lower elevation flood plains.

The population of Sumba is about 685 000 people, and the town of Waingapu on the north-eastern coast about 52 000 people. Waingapu is the administrative centre and main entry point for the island. To reach it you board one of the international flights to the island of Bali. At Denpasar Airport you stand in long queues next to pale travellers. Crowds of Balinese ask to carry your luggage, for tips. A domestic flight from Denpasar takes you south-east to Sumba, back towards Australia. If your plane fails to show up for the scheduled flight, and this often happens, you catch a ferry and cut your losses. This takes two or three days out of your schedule and is factored into your planning.

Remote Sumba is quiet, thinly populated and mainly Christian. Barn Owls live in the steeples and bell towers of Catholic and Protestant churches there. Taxonomists in Australia and elsewhere use feathers to analyse DNA relationships between species, and these bell towers held feathers of roosting and nesting birds, as well as pellets and prey remains.

In a pristine white-tiled Lutheran church, a church member removed the manhole cover over the lobby so we could look up through the wooden struts and girders into the bell tower. The white face of a Barn Owl stared down. Barn Owls are one of the world's most widely distributed land birds, they live on every continent except Antarctica and on many islands, so researchers study them in Europe, Australia, Asia and many of the world's islands. David Hollands said that Barn Owls have an aversion to cold, so they don't occur in the far north of Eurasia and North America. In many parts of the world they nest in barns and church towers like this one in Waingapu, but not in Australia. In Australia they nest almost solely in tree hollows. This difference is a mystery.

The delicate beauty of a Barn Owl always startles you (Figure 42.1). Clean and white with gold feathers on its back, a heart-shaped gold rim around its face, with clear black eyes.

Inside the ceiling of the church lay gold and white primary feathers dropped by the Barn Owl and its mate. The ladder was shaky and it threatened to come loose from the wall. The heat was fierce up in the tower; further up lay more feathers and castings. At the top of the ladder there was a hint of breeze coming through slats in the bell tower. The Barn Owls hissed and wheezed, swayed back and forth, behaviours that characterise *Tyto* but not *Ninox*, *Bubo* or other owls.

Barn Owls are rodent specialists in Europe and North America, but little is known about their prey in tropical regions. *Tyto* castings differ from other owl castings because they are covered with a black glaze. From the castings caught on wooden beams inside the ceiling space over the church lobby, we could see what this pair ate – Tree Sparrows, some kind of rat, probably introduced Black Rats or the endemic Pacific Rat and fruit bats. At another nest on the nearby island of Roti, they ate mainly fruit bats. In the Galapagos Islands Barn Owls ate many introduced House Mice and Black Rats, but a study from Argentina and another from Hispaniola said they also took bats.

Motorcycles buzzed along the main road below, and a knot of townspeople had gathered in the churchyard to watch. In 1997, a speaker at the Winnipeg owl conference had painted a grim picture about public knowledge of owls in certain Third World countries. In Costa Rica in Central America, and Malawi in Africa,

Figure 42.1 Barn Owl in the church tower on Sumba.

strong negative superstitions led to unnecessary killing of owls. People, particularly old people, believed that hearing or seeing an owl brought them closer to death or catastrophe. Costa Rican beliefs, the speaker said, may have emanated from Africa, the origin of black Costa Ricans who carried the beliefs with them. The authors thought one possible explanation for the persistence of the Malawian myths was that owls roosted, called from and hunted in graveyards. The Malawians deforested areas around their own villages so graveyards stood as the only remaining woodlands. However, these beliefs were common throughout Indonesia. In the Sumba countryside, people buried their relatives in front yards, close by so they could talk to and visit them. Barn Owls on Sumba were called *Burung hantu* 'ghost birds', and they were a sign of bad spirits and coming death or illness, partially because of their white form. Even in farming villages, where *Tyto* owls lived in the fields away from ancestors' graves in front yards, owls were connected to ghosts.

We found Barn Owls at night in the deep canyons that cut through the Sumba plateau. At night Barn Owls were the only night birds we saw, flashing white in the headlights, flying over the road or perched on fence posts in the rice fields. Along the winding road between Waingapu and Lewa, a Barn Owl lay dead on the road with a string tied to its leg. We stopped and two boys emerged from the roadside

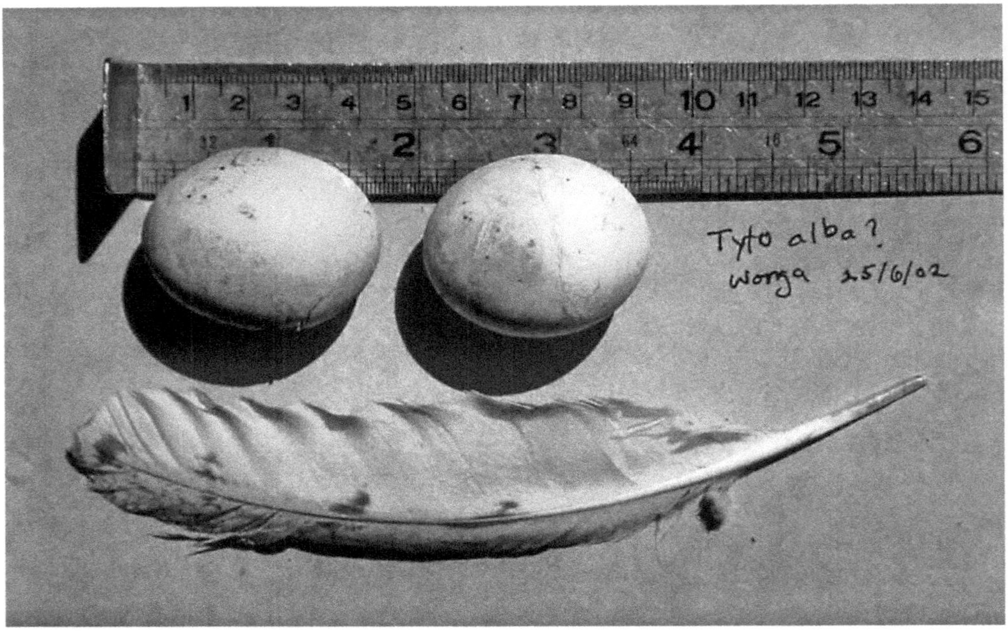

Figure 42.2 Barn Owl eggs from a nest on Sumba.

bushes and talked to us through an interpreter. The boys had killed the owl with slingshots, at its day roost in a limestone cliff, and slung the body on the road to watch it get run over. Killing owls was a common sport on Sumba.

On 24 June 2002, 1600 h, near Wanga, we found an old nest in the attic of an abandoned brick-walled building formerly housing a water-pump. Three white eggs, and primary feathers, assumed to be from the incubating female, were resting on debris, castings and rotting wood (Figure 42.2). One egg was broken; the two intact eggs measured 41.9 × 32.9 and 41.1 × 33.4 mm, similar in size to other Barn Owl eggs.

Measurements in millimetres of this freshly killed owl were: wing 275, tail 107.4, mid-toe with talon 3.36, mid-toe talon 1.28, inner talon 1.61, hind talon 1.86, culmen without cere 230, wing formula 9/8/10/7 from outside and it weighed 385 g. These measurements fit measurements for *Tyto alba delicatula* given by König *et al.* (1999) and Higgins (1999), but further analysis is needed before the taxon on Sumba is treated as a separate species or a subspecies of *Tyto alba* or *delicatula*.

Molecular analysis of Barn Owls by Wink *et al.* (2008), including feathers from the dead owl on the road on Sumba, and moulted feathers from the church tower in Waingapu, placed Barn Owls from Sumba into *Tyto delicatula*, with Barn Owls from Australia. However, the authors remarked that *Tyto delicatula sumbaensis* are distant enough from the others to probably merit species status. Christidis and

Boles (2008) considered all Barn Owls from southern Asia through Australia and the Pacific to be one form, *Tyto javanica*, and used the common name Eastern Barn Owl, although alternative common English names, they said, should be considered. The taxonomy of the world's Barn Owls is unclear.

At one point we saw a Grass Owl fly over the road, very much like a Barn Owl but with longer legs. Main texts like Konig *et al.* (1999) and Higgins (1999) said Grass Owls did not exist on Sumba, but as mentioned above, ornithology in Wallacea is poorly known. So far there are no feathers from the Sumba Grass Owl, so taxonomists do not know if it is the mainland Asian and Australian form, the Eastern Grass Owl *Tyto longimembris longimembris*, or one of its subspecies from nearby islands such as the Philippines. It could be a new species.

43
Sumba Boobook

The strip of mangrove forest along the east coast of Sumba held a few saltwater crocodiles. Local farmers warned us about one territorial male who, they said, was bad-tempered. At night, down near the tidal edge, he grumbled as we returned through tall grass to the car. An impoverished corn farmer said that local boys teased this crocodile and this teasing made him irritable. On the wall of the corn farmer's bamboo hut were the dried sex organs of another male crocodile that had lived in the mangrove swamp until, the farmer said, he had dived into a pool and killed it with a knife. He promised to dive into the lagoon and kill this one if we photographed him doing it. We declined.

This patch of tall forest near the swamp held a pair of Sumba Boobooks *Ninox rudolfi*, a hawk-owl cousin of the Southern Boobook. They hunted through forest, swamp and nearby grassland, and we fought oppressive humidity each night trying to see them.

After two weeks of study we guessed, from patterns of calling, that two pairs lived in this swamp. The species had not been photographed or videoed before, and clear stills and video could help settle arguments about the owl's taxonomy and character. Video of calling adults would contain audio data to convert into sonograms, visual pictures of the owls' calls that could be analysed. Some writers thought that Sumba Boobooks and Southern Boobooks were the same species, and comparing sonograms would sort this out. If they had the same calls, they were the same species. If they had different calls, they were different species.

What is a species? Following the Biological Species Concept, a full species is said to be a reproductive community that has evolved different patterns of behaviour from members of another reproductive community. Behaviour is the key word here because two species can look alike and even live in the same place. The line separating the two species is not always clear, but researchers can discriminate owl species by their calls, their territorial songs. An owl's call can be like a DNA fingerprint. The call is inherited and unlike the calls of some other birds such as songbirds, it has little geographic variation in dialect, so the call tends to be similar across the owl's range.

The Sumba Boobook is an endemic, from only one island. Only three specimens exist, all shot before 1900 and all in German museums. Some writers said the male was larger, some said the female was larger. Writers were uncertain about whether the short cough-like call recorded by earlier researchers was the real territorial song or just a variation of the Por call used by Southern Boobooks (Chapter 19).

Surveys in 1989 and 1992 recorded only small numbers of Sumba Boobooks in five places, in monsoon and rain forest, both primary and secondary. They seemed to use forest at high elevation as well as forest along the coast, like this swamp near Wanga. Because of the familiar story of habitat loss, Sumba Boobooks were said to live in diminishing forest and BirdLife International listed them as Vulnerable.

Many locals did not trust white people, a consideration for researchers. Locals still believed that whites captured Sumbanese women and children for slavery, as Dutch colonists had in the past. Some children screamed and ran away from us. One evening, as we walked through dark forest along a river near Waingapu, Sue played a recording of a Sumba Boobook to elicit a response. We passed a woman and her daughter who saw us, screamed, and ran back along the muddy path to their village. Ten minutes later, a village leader cycled down the path to investigate; our guide talked to him and calmed him.

Each afternoon, we travelled along the coast road from Waingapu, past villages, hunting kestrels, farmers working open fields, school children in clean uniforms and cinder block houses with the owner's relatives planted in concrete graves in the front yard. We would stop at different woodlands after dark and check for owls. It is best to explore a woodland or grassland on foot in daylight so you can better navigate the same woodland at night after owls emerge. One late afternoon, while searching open woodland along a white sandy beach near the community of Wanga, we accidentally stumbled into a fishing camp, economically tied to the sea (Figure 43.1). Washed clothing lay drying over bushes in the sun, and fishing nets hung in neat folds over tree limbs. The fisherman had ingeniously cut circles from the soles of rubber thongs washed up along the beach as floats for these nets.

A group of young men roared up on motorbikes. Our guide took the young men behind one of the huts and there was fierce talk for 10 minutes, as if a fight might erupt.

Figure 43.1 Fishing is an important industry for villagers in Indonesia.

They offered us betel nut, reddish-brown, about the size of a thumbnail. It looked and tasted like tree bark. It gives a mild buzz before turning your teeth red. Sue refused hers, one of the villagers became agitated and our guide spoke on our behalf. The villagers relaxed, we left to search more woodland.

In spite of their mistrust of white people, most farmers were kind and offered help (Figure 43.2). They have little food and few possessions, but will invite you into their low huts or wooden houses, spread out mats to sit on and, through an interpreter, offer local knowledge about owls and other animals.

After dark, fruit bats left trees and flew across openings. Churring insects buzzed and geckos made loud, croaky *'ge-cko'* calls through the forest. Our first Sumba Boobook flew into a tree above us and called. It had striking large white spots over its dark head, dramatic rufous bars across its breast and liquid black

Figure 43.2 Villager on Sumba selling freshly cooked corn.

eyes. The SLR film camera and 200 mm portrait lens, with a 2-times multiplier and a flash, was too weak to reach owls in the treetops. We needed to draw them down closer using a broadcast of their own voice, recorded by video in the dark and then copied to audiotape. Often the Sumba Boobooks didn't respond to the tape, or they might drift in, ghost-like, then disappear.

Soupy mist on the mangrove shore fogged the space between the lens and glass filter screwed onto the front. We removed the filter and dried the front of the lens with a cloth, but it would fog up again and rubbing the front element of the lens too often in a gritty swamp at night could ruin it. The black lens barrel was coated with silvery dew thick enough to write our names in.

When the humidity rose too high, the electronics in the video camera shut down. But on some nights the swamp air was less humid and the lens stayed clear. Sue would play the tape; our guide and I would rush through woodland carrying the camera and tripod towards the owl. The owls continued towards Sue and came over our heads. We stood between Sue and the owl snapping photos.

The female Sumba Boobook gave a food-begging call to the male, similar to the food-begging call of female Southern Boobooks. This was no surprise; the two species were cousins. Sumba Boobooks also made a sort of chuckling squeal that we termed 'monkey calls'. At first locals told us these *were* monkeys, not owls, and monkeys did live in the same forest. But we saw the owls making these calls (Figure

Figure 43.3 Sumba Boobooks occur on only one island.

43.3). We had never heard Southern Boobooks make such a call, or make the *cluck-cluck-cluck* call that male and female Sumba Boobooks made.

On video the larger in the pair, to our eyes, gave deeper calls, so these probably were males, as suggested by Konig *et al.* (1999) *contra Handbook of Birds of the World.*

Measurements of the Sumba Boobook differed from those of Southern Boobooks. Sumba Boobooks were similar in size to the Southern Boobook subspecies *N. novaeseelandiae ocellata* said to occur in northern Australia and on the nearby island of Savu. *N. novaeseelandiae ocellata* are smaller than Boobooks near Canberra; however, Sumba Boobooks had noticeably shorter toes than either Boobook subspecies (Table 43.1). They appeared to be much slower moving when rousing, scratching or turning around on the perch than were Southern Boobooks. Though smaller, they moved like a much larger owl, like a Powerful Owl or a Great Horned Owl. Sumba Boobooks probably do not catch many flying birds or insects as do Southern Boobooks, a quick-moving, quick-flying species that hunts in the manner of accipiters. One Sumba Boobook attacked a rat held in a cage by villagers and knocked the cage over, so they probably take rodents, but also reptiles such as geckos, a common animal 'churring' all night in the owl's habitat. Raptors that capture reptiles or small rodents often have proportionately shorter toes than species that capture birds or aerial insects (Johnsgard 1990) and this is borne out in the comparative measurements in Table 43.1.

Table 43.1 Measurements of Sumba Boobook and two subspecies of Southern Boobooks in mm; weight in grams. (1) one unsexed Sumba Boobook *N. rudolfi*, and means from (2) Southern Boobook *N. novaeseelandiae ocellata* (a– males n= 20–22, b– females n= 24–26; from Higgins 1999 p. 870) and (3) *N. boobook boobook*, Canberra (a– males n=9, b– females n=5; Olsen and Trost, unpublished)

	Wing	Tail	Weight	Culmen	Mid-toe	Tarsus
(1)	231	133	222	160	225	392
(2a)	222	124	216	159	266	410
(2b)	227	125	234	168	267	410
(3a)	240	142	272	164	278	519
(3b)	246	143	327	169	294	516

Most *Ninox* owls have a two-note call like the European Cuckoo, but the song of the Sumba Boobook was a continuous *cluck-cluck-cluck-cluck-cluck* with each note repeated about 2.3 times *per second*, very different from the call of the Southern Boobook's double-note call repeated about every 3 seconds. Locals told us that the *cluck-cluck-cluck* calls of the Sumba Boobook increased before the wet, when pigs moved into certain swampy areas near Wanga so they called this owl 'Katuawe', which means 'leader of the pigs'.

Sumba Boobooks and Southern Boobooks are relatives, but not the same species. Given the video and audio evidence we proposed a new common name for the Sumba Boobook, the 'Sumba Hawk-Owl'.

Recorded vocalisations need formal analysis. In the next chapter we compare sonograms from these two 'Boobook' species, and a newly discovered owl.

44

New species

Religion played a part in our search for Barn Owls in church towers on Sumba, and it played a part in our search for the 'mystery owl'. Many people on Sumba are Christian, but their Christian beliefs overlay traditional ancestor worship and animist *marapu* religion. Spirits of the dead look after the welfare of the village and ensure good harvests and these spirits need ritual appeasement. Tribal enemies on Sumba used owls to steal the spirits of family members, so locals were suspicious when we came looking for owls.

Some guides refused to enter woodlands because of spirits, and some drivers refused to wait near woodlands as we searched for owls. They drove off. To some villagers Sue looked like a spirit. When we arrived at one farm and Sue, with her copper-blonde hair, got out of the car, two small children playing in the yard scrambled under the house, crying. Our guide told us about a blonde 'European' woman (this could mean Australian, British, French or any number of nationalities) who loved the broad grasslands on the Sumba plateau. The woman took to wandering by herself and drifted too close to a village. The village men killed her, thinking she was a spirit. They felt bad about their mistake, and Sumba courts let them off because the mistake was understandable.

West of Waingapu you wind up a steep grade, slowing at many switchbacks, grinding in low gear, stuck behind trucks carrying green produce and farm labourers, onto the grassy plateau, elevated and cooler than Waingapu. Like an open park, it is cut by limestone canyons and dotted with patches of woodland, narrow lanes and freshwater lakes.

At dusk we parked the vehicle at the foot of a heavily timbered mountain, loaded video and still cameras, water and food into a pack, and followed our guide up a steep muddy grade to a grassy clearing on top of a hill. Fireflies floated in low bushes and giant insects buzzed past our ears and swarmed through holes in the canopy. Native pigeons came in to roost for the night, bowed and grunted like pigs.

Our guide stopped at some jagged pieces of limestone laid against the trunk of a deciduous tree. He whispered, 'At this place hunters pray to the gods for good hunting in this forest'. Bird hunters rested at the bone-white limestone altar and spoke to their ancestors.

We waited, listening to the forest.

Sue pulled a portable cassette player from her pack. We stood huddled in the high grove and broadcast the voice of the 'mystery owl', the single repeated 'hoot' that Mark Linsley from England had recorded on Sumba and kindly given to us. It sounded like a child blowing the same high note on a flute every three seconds.

Some twitchers and ornithologists said the 'mystery owl' was a scops owl, the genus of tiny owls with ear tufts and cat faces. As mentioned in Chapter 41, species of *Otus*, or scops owls, occur in North and South America, Eurasia and most of the world except Australia where *Ninox* Hawk-Owls dominate.

Woodcarvings of owls sold in Balinese gift shops, just west of Wallace's line, have little cat ears – scops or eagle owls. Ornithologists who glimpsed the 'mystery owl' believed it was a scops owl, like the nearby Wallace's Scops Owl on Flores, named for Alfred Wallace. Experts had identified three species of scops owl from Flores, the island you can see through binoculars off the north Sumba coast. Maybe scops owls had drifted across from Flores and the 'mystery owl' was taxonomically related to Eurasian owls, not Australian *Ninox* owls.

Some writers dismissed these reports as a misidentification of the endemic Sumba Boobook, the owl we had seen and photographed near Wanga, a hawk-owl with dramatic rufous bars across its front. Other writers, who saw 'ears' on the little owls in their spotlights on Sumba, said the bird was certainly the Flores Scops Owl. The territorial song of the Flores Scops Owl was unknown, but the song of the 'mystery' owl was unlike any known scops owl. One ornithologist, Ben King, played the song of the 'mystery' Sumba owl over on Flores, close to where the Flores Scops Owl had recently been rediscovered, but he failed to elicit a response.

In the grove of trees on the Sumba mountainside we broadcast the call of the 'mystery owl' where locals had seen it. In late December, the edge of the wet season, the air was thick and we could smell damp leaves and wet soil. No owls replied. After standing in the dark for half an hour, listening to croaking geckos, buzzing insects, and grunting pigeons, we walked back downhill, sliding, arms outstretched to balance in the mud. We packed the vehicle and drove the twisting Lewa road to Kilometre 49, where bird-watchers reported they had heard the 'mystery owl'.

Figure 44.1 Walking over grassland on Sumba to the arm of forest where the new species was discovered.

We left the vehicle and, by moonlight and torch, hiked 800 m along a ridge, over scorched grassland, around blocks of bone-coloured stone the size of houses, to an isolated arm of forest. Inside the forest moonlight was blocked and it was darker as we found our way between high trees. Waist-high spikes of limestone poked up through bushes; vines wrapped around tree trunks. We took care not to trip and cut an elbow or knee, open a wound that would bleed, draw insects and fester in tropical air.

From a grassy clearing inside the forest at 1830 hours Sue broadcast the taped voice of the 'mystery owl'. There was no response. We didn't talk, expecting to be disappointed again. We heard croaky '*ge-cko*' calls, churring night insects, coos of settling pigeons and saw slow-flapping fruit bats in the night sky, but no owls.

After 10 minutes two owls called back. We tracked them in the dark as they drifted downhill from tree to tree, then to the forest edge. They hid behind leaves and branches in the canopy, so we couldn't see them by torchlight.

While Sue broadcast the call down in the clearing, our guide and I entered the forest with the video camera, pulling aside vines between spikes of limestone, trying to glimpse the owls and video them before the calling stopped. They were shy. Each time we drew one close enough to light it with the torch, both owls slipped behind leaves or flew to another tree.

Figure 44.2 Because it tended to conceal itself in the forest canopy, the 'mystery owl' on Sumba had not been photographed. This is the first.

After half an hour, one sat directly overhead, calling. This owl was reddish, the size of a quail, had loose feathers and bright staring eyes, like yellow buttons on the front of its face (Figure 44.2). The owl had no ear tufts, as scops owls are supposed to have, and we realised that it may be a new species of hawk-owl, a cousin of Boobooks in Canberra.

Over the next two weeks we searched for more. Two other pairs lived in adjacent forest, one accompanied by a rufous-coloured juvenile, head-winding as juvenile Boobooks did in Canberra. The juvenile flew in the canopy with the adults and trilled. We gathered data on calls and behaviour, and videoed what we could. On most nights Sumba Boobooks called in the same forest with their rapid *cluck-cluck-cluck* calls.

On 30 December 2001 a guide in Waingapu showed us a specimen of the 'mystery owl' that a bird hunter had shot in a patch of forest some 4 km east of our study patch. We measured and photographed the specimen, then returned that night to Kilometre 49. All three pairs were there and calling (Figure 44.3).

When Professor Michael Wink at the University of Heidelberg in Germany received feathers from the specimen, he analysed nuclear marker genes to determine which taxonomic group the owl belonged to. It was not a scops owl, but a *Ninox*, a hawk-owl like the Sumba and Southern Boobooks. This little owl

Figure 44.3 Pair of Little Sumba Hawk-Owls high in the forest canopy.

weighed about 90 g, a third the weight of a Southern Boobook, and differed by 8.2 per cent nucleotide substitutions.

This owl differed in other ways from known *Ninox* owls. It was half the size of a Sumba Boobook, with different markings and a different call. As mentioned above, calls are important in owl taxonomy. The song of this new owl was a single note, a monosyllabic 'hoot' repeated every 3 seconds. It was like a child monotonously blowing a single note on a plastic flute, unlike any *Ninox* song known at the time, including the larger Sumba Boobook in the same high forest with its rapid, continuous *cluck-cluck-cluck*.

From audio recordings taken from video, Michael Smith at the University of New South Wales compared three call properties from the owls: (1) dominant frequency, (2) call duration and (3) call rate (see Table 44.1 and Figures 44.4(a), (b) and (c)).

In our *Emu* paper we proposed a common name, Little Sumba Hawk-Owl after its island home, and gave it a scientific name, *Ninox sumbaensis*, the 'Hawk-Owl from Sumba'.

The owl tally for Sumba now stood at four species – two *Ninox* or hawk-owls, and two *Tyto* or barn owls, though scientists have no specimen of the Grass Owl

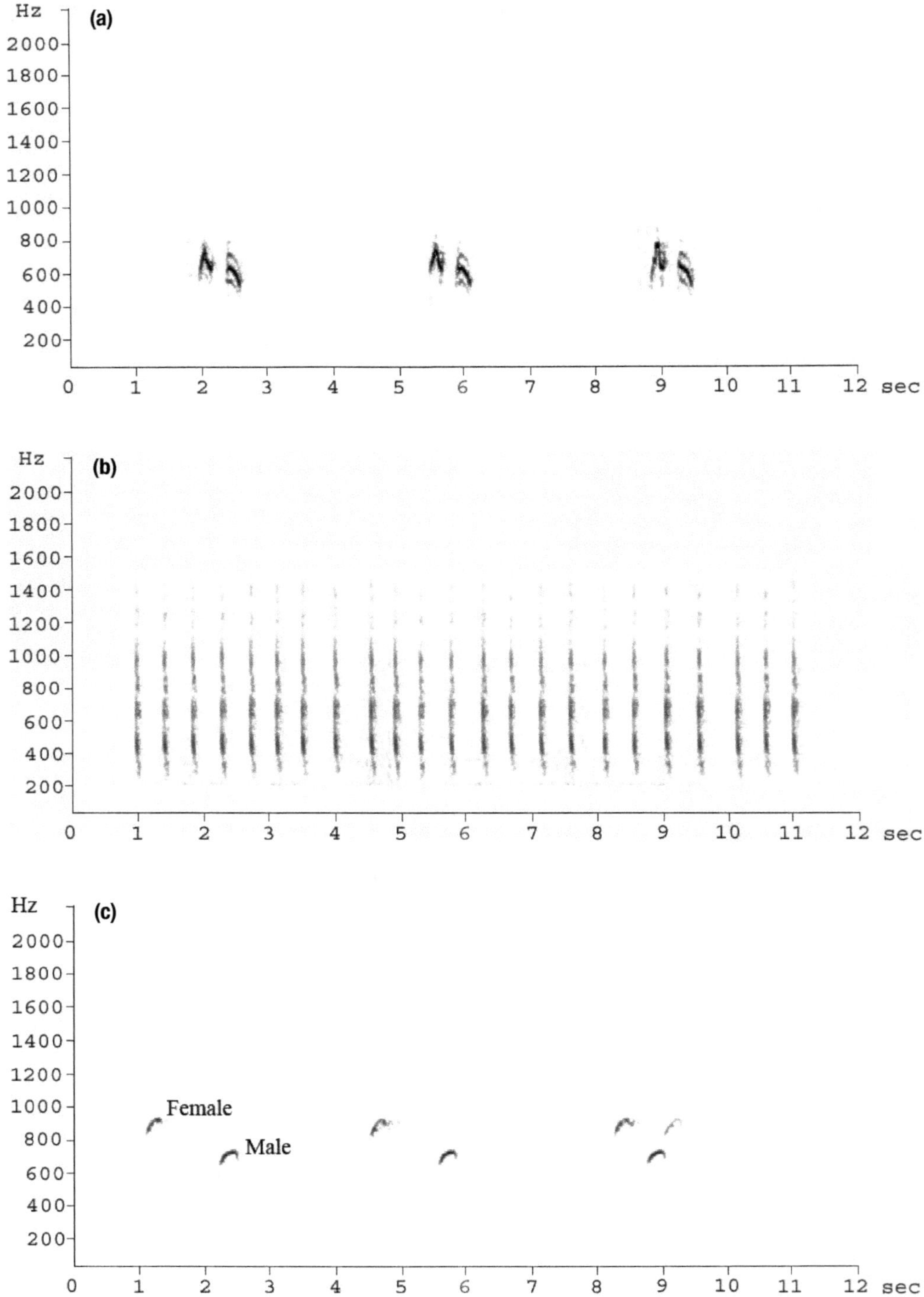

Figure 44.4 Sonograms for (a) *Ninox novaeseelandiae*, (b) *Ninox rudolfi* and (c) *Ninox sumbaensis*.

Table 44.1 Call properties for three owl species – Mean and ± standard deviation from the audio track of video recorded Sumba Boobook *Ninox rudolfi*, Little Sumba Hawk-Owl *Ninox sumbaensis* and Southern Boobook *Ninox novaeseelandiae* (after Olsen *et al.* 2002c)

Species	Dominant frequency (Hz)	Call duration (ms)	Note duration 1st note (ms)	Note duration 2nd note (ms)	Call rate (calls/min)
Ninox rudolfi	714.8 ± 23.6	38.6 ± 2.6	–	–	138.0
Ninox sumbaensis	871.8 ± 21.5	383.0 ± 26.1	–	–	24.0
Ninox novaeseelandiae	846.8 ± 35.1	609.7 ± 49.2	222.0 ± 21.9	222.0 ± 21.9	18.0

and do not know what species it is. Researchers have found no scops or eagle owls on Sumba, but this can change with another visit. Wallacea is unexplored ground.

The conservation status of the Little Sumba Hawk-Owl is unknown. With too little forest on its native island, we hoped it would not go the way of the Laughing Owl in New Zealand or the Lord Howe and Norfolk Island Boobooks, but continue to give its sharp fluted call in high forests on Sumba.

45
Eviction

Disappearing woodland in Canberra gave some indication of pressures faced by Sumba owls. Humans degrade patches of forest, chase owls from their breeding sites and diminish food. For Little Sumba Hawk-Owls, an endemic island species, there was no place to go.

Habitat degradation is an abstract concept. We hear the term so often it loses its emotional impact. It is not an abstract to animals. How do owls cope? How do owls behave when prime habitat is taken away?

Along the new roadway in Canberra (see Chapter 35), loss of trees had increased the conflict (Duelling) between neighbouring male Boobooks; three of five pairs left, leaving two. Commensurate with loss of this habitat, a new pair showed up in suburban Cook where none had bred before. They tried to breed in sub-prime habitat. There was plenty of food – introduced mice, rats, sparrows and mynas – but the Boobooks had another problem. Too many big trees felled near the city left too few nest hollows. Animals like Brushtail Possums and Sulphur-crested Cockatoos were increasing and fighting over hollows that were left.

In 2005 a pair of Boobooks laid eggs in an Apple Box in a small park along a bike path. Houses lined both sides of the park. There were dozens of resident Common Brushtail Possums, in plague numbers because humans fed them. The possums left their tree hollows at night and some foraged on eucalypt flowers in the canopy. Others visited backyards and begged for food. That year some possums tried to enter the Boobook's nest hollow, and sometimes succeeded. Sulphur-crested Cockatoos stopped Boobooks from using one hollow and sometimes harassed them. As well, this pair of Boobooks faced increasing levels of Duelling

Figure 45.1 Each night the Boobooks attacked Brushtail Possums in the nest tree.

from a male on their eastern border, again related to trees lost when the new road pushed through.

Each night the female Boobook came out hungry, begged from the male and watched for enemy possums. She eyed each possum as it came out of its hollow, puffed up and clacked her beak. If one tried to climb past her nest hollow into the tree canopy to feed on eucalypt flowers, she attacked it. Animosity and tension ruled each night. She attacked possums, they dodged her talons and beak. However, a possum entered the nest hollow one night and apparently ate or smashed the eggs. The Boobooks failed in 2005 and the female drifted off to winter elsewhere. The male remained.

In 2006 the same male and a different female tried again, in a tree close by. The new female hunted insects near the nest until possums emerged. She attacked, clacking her beak, slamming her talons into their heads and backs (Figure 45.1). Each night, possums growled, Boobooks swooped, possums ran along branches. The air crackled with hostility.

Eight possums and at least one Sugar Glider denned in the Boobooks' nest tree. To escape attacking Boobooks, the possums ran along fixed routes to an extended branch at the top of the tree. They would leap to an adjoining tree and flee, but the owls even attacked possums in this adjoining tree.

One possum would stay back, in its hollow close to the Boobook nest. He seemed intent on trouble. Anthropomorphically, he seemed to have cunning about him, maliciousness; compared to other possums, he was less frightened of the owls. If he saw a chance, he probably would enter the Boobook nest and smash the eggs, as happened in 2005. The female or male Boobook always drove him away just in time, but this possum was not cowed.

In the first hour after dark the female Boobook often left the nest hollow to stretch or hunt insects, and the male hunted away from the nest, then returned with a mouse, centipede, moth or cricket. If he returned and the possums were in the nest tree, he delivered food to the female then attacked them, clacking his beak. He often struck them, but the female did most of the defence. She knocked possums out of the tree a few times. Crawling, head-down the trunk, possums could not grip as well. If she hit them sideways in the hindquarters, they lost their grip and fell. Possums are tough customers, and they puffed up and landed with a thud. The tumbles did not kill any.

On the night of 14 November 2006, the female left her nest tree and eggs at 20:22. She had a habit of caching food in the top of a tree over the bike path, then retrieving it at dusk. She was absent for eight minutes. The 'bold' male possum stayed in his hollow; he seemed intent on trouble. He climbed up the branch and entered the Boobook nest hollow at 20:30. The female flew back from her food cache and stood in the entrance, beak-clacking, puffing up to look big as possible, all too late. The possum was inside.

Boobooks weigh about 300 g, possums about 3000 g, 10 times the weight of the owls. Owls dominate the possums in the open, possums dominate the owls in tight spaces, in alley fights.

The possum growled from inside the hollow. We held a torch on the tree and glassed the hollow entrance, the possum inside, the agitated female outside, unable to use aerial strikes. After 50 minutes the possum emerged, his feet and mouth damp. He had eaten the eggs.

The possum sat facing out of the nest hollow cleaning wet egg off his feet and mouth. He growled at the female when she came too close, as if taunting her. He left at 21:20 and the female disappeared inside the nest hollow until at least 22:00 when we left.

There was no sign of the male Boobook. The nest had failed two years running.

Over the next few days the female took food deliveries from the male and returned to the nest hollow. Maybe she would lay another clutch, and possums would eat that one too.

On three consecutive nights of one-hour observations, 17, 18 and 19 November, the Boobooks hit possums 16 times. On 17 November the female knocked a possum out of the tree. It puffed up, spread its feet and landed with a thud on the grass. She attacked it as it ran along the ground.

The female attacked another possum and the male hit one trying to move down the trunk to the ground. Each night there was beak-clacking and growled warnings, like a battlefield.

On 3 December the female brought five insects to the nest hollow and disappeared inside each time; the male brought in a mouse. The female remained inside. With the female in the hollow the male launched four attacks on possums, three of them hits. There was trilling inside the hollow. Apparently, the possum had not eaten all the eggs.

Over the next few nights, the adults delivered food, and slammed into possums. On 14 December the 'bold' possum went close to the nest hollow and looked in, but did not enter.

Battles continued through Christmas. On the nights of 15 to 26 December we noted 17 attacks and 11 hits on possums.

On 27 December, the female left the nest at 20:41. At 20:45 the juvenile sat looking out of the nest hollow, our first view of him. The female made five prey deliveries, the male one at 21:08. Then both adults left.

At 21:52 'bold' possum poked his nose out of the hollow below the nest. He looked around. Nothing happened. The Boobooks were gone, looking for prey. The possum crawled along a branch to the trunk, up to the edge of the nest. The parents did not appear. He moved to the lip of the nest hollow.

The juvenile Boobook slid back into the hollow clacking its beak, trying to get away from the possum, but he was trapped. Thirty seconds later the possum entered the nest hollow and we believed he would kill the owlet. There was beak-clacking, some rustling inside, then it went quiet.

The owlet backed out of the nest hollow with its wings spread and sat on the edge of the nest, trilling for his parents. They did not show.

The possum came out after the nestling, smacking the nestling with one paw. The owlet crawled up higher on the limb, beak-snapping at the possum, who backed down into the hollow and sat with his head poking out. Positioned here, the possum could win any fight against attacking Boobooks.

At 22:30 the possum moved out of the hollow towards the nestling again. He batted at the juvenile, trying to grab him, or maybe trying to knock him from the tree. He forced the juvenile off the edge of the hollow and up onto another fat, slippery limb. The possum crawled towards the owlet and whacked him again. The owlet fell and disappeared from the circle of torchlight.

In the darkness we could not see where he landed, or if he was hurt.

We hunted around the area but could not find him. Young owls occasionally slip when flapping their wings at the nest entrance, practising flight as they grip the edge of the hollow. On the ground they often sit motionless. We broadened the search under bushes, because dogs or cats would kill him in the morning. We found him under a bush and took him home, arranged a place for him in the bathroom, fed and banded him (Figure 45.2).

Figure 45.2 After it was ejected, we kept the nestling for one night.

28 December 2006

We left him in a darkened cardboard box for the day. At 20:44 the female sat in the nest entrance. I climbed an adjacent tree with the owlet, hooked his feet to a branch, waited until he had a good grip and left him. The female Bray called. The male flew over and sat next to him. The male left, caught a moth and tried to feed the nestling. The nestling trilled but would not take any food, apparently too nervous or excited (Figure 45.3). Both adults circled close to the nestling in the tree, as if trying to shepherd him back to the hollow. The juvenile flew weakly across to the nest tree, clambered up some long branches and entered the nest.

He was home.

29 December 2006

At 20:32 the female sat in the nest hollow entrance. At 20:44 the juvenile peered out sitting next to her. She left at 20:47 and brought back an insect. The juvenile stayed in the hollow during our observation and no possums attacked him.

Figure 45.3 When the nestling was released, the adult male caught a moth and tried to feed him, but the nestling refused.

On 30 December there was a major hailstorm. Rain continued through the night.

31 December 2006

Severe thunderstorm, followed by more heavy rain. When the rain lightened, we visited the nest. The juvenile had left the nest tree and was calling from a nearby tree that possums often used to escape attacking Boobooks. Both adult Boobooks stayed close, as if protecting the juvenile. He trilled with an adult sitting next to him; we left at 21:45.

1 January 2007

The owls were gone. Possums may have interfered again, or the juvenile may have died in the rain. He was fluffy, not waterproof yet. On 1 and 2 January we searched bushland, horse paddocks, suburban streets and gardens but failed to locate them.

We looked for several more nights, then we gave up.

2 February 2007

A month later we found the juvenile in a small park 500 m from the nest. The park had good leafy trees for catching insects and birds, but no nest hollows and few possums, a suitable place for a juvenile Boobook to serve an apprenticeship. He was trilling and following the adults. The Boobooks apparently left the nest area on the night after he fledged, maybe to escape the possums and hostility.

Two nights later we observed for one hour. He flew well, trilled occasionally and hunted insects on his own. In spite of the loss of nest trees, conflict with other owls, attacks from possums and cockatoos, they had raised a single nestling to independence.

They made it.

Appendix A: Australian owls

This section is not a comprehensive summary of each Australian owl. Rather, it gives a brief description of each of nine species and points to questions for further research. For comprehensive descriptions of the biology of each species consult Fleay (1968), Hollands (1991), Higgins (1999), König *et al.* (1999), Garnett and Crowley (2000), Garnett *et al.* (2011), Newton *et al.* (2002) or Debus (2009).

The world's owls are divided into two families: (1) the Tytonidae or 'barn owls', and (2) everything else fits into Strigidae or 'typical owls'. The two families show similarities in breeding and methods of hunting; owls are similar around the world in behaviour and appearance. Many of the differences between the two families are anatomical. Tytonidae owls have inner and central toes of about the same length, while Strigidae have an inner toe that is shorter than the middle toe. The claw of the central toe in Tytonidae is serrated on the underside, in Strigidae it is not. The skull of Tytonidae appears to be long and pinched, in Strigidae comparatively short. The shape of the sternum (breastbone) is different in the two families. In Tytonidae the sharp edge of the sternum, the carina, is broad and narrows towards the abdomen, while in Strigidae it is narrow at the upper part and broadens towards the abdomen. In Tytonidae the sternum has two shallow notches near the abdomen, in Strigidae there are two deep emarginations on each side (Figure 2.3).

Tytonidae are said to be distinguished from Strigidae by a sharply defined facial disc, typically heart-shaped (Higgins 1999; Figure 2.1) but, to my eye, the differences are not that clear between some Tytonidae species like the Sooty Owl in Australia and Northern Hemisphere Strigidae like the Boreal Owl *Aegolius funereus*. Even some individual Southern Boobooks have a fairly well-defined facial disc.

Tytonidae is said to have only one genus, *Tyto*, though recent taxonomists add *Phodilus*, the Bay Owls. Strigidae has about 24 genera, give or take a few depending on taxonomic treatment. Though even small countries like England have six owl species across five genera, the Australian mainland has only eight species in two genera, *Tyto* and *Ninox*. In Australia these two genera fill the niches that are often filled by a number of genera elsewhere in the world.

Tyto owls

The genus name *Tyto* is derived from the Greek for a type of owl, probably named for its cries. *Tyto* can be divided into four groups: Sooty Owls, Masked Owls, Barn Owls and Grass Owls (Christidis and Boles 2008). All four are represented in Australia, the Barn and Grass Owls that prefer open land, and the Sooty and Masked Owls that prefer forested land.

Eastern Barn Owl *Tyto javanica*

The species name *javanica* is from Gmelin 1788 for Barn Owls in southern Asia. Wink *et al.* (2004) placed Australasian forms, through Australia and the Pacific into this species, separated from *T. alba*, and Christidis and Boles (2008) followed this, though König and Weick (2008) used *Tyto delicatula*, and argued that *javanica* belongs with *alba*.

Field recognition: A medium-sized owl that often appears white in the field at night.

This slim *Tyto* has slender legs and the lower parts of its tarsi are sparsely feathered. Length 29–38 cm; wingspan 70–100 cm; weight 310–360 g. The **adult** has rather small dark eyes compared to many owls, set in a white, heart-shaped facial disc with a narrow dark buff rim and a narrow dark tear-drop below each eye. On the upperparts it is grey, washed buff and finely spotted black and white; on the underparts it is white with fine dark spots. The bill and cere are pinkish-white, and the feet are cream or light brown to dull pink. The **juvenile** has white down on the head and underparts, which is soon shed (Higgins 1999; Debus 2009).

Subspecies: Higgins (1999) recognised one subspecies of Barn Owl in Australia *Tyto alba delicatula*. Wink *et al.* (2004) analysed cytochrome b from feathers dropped by Barn Owls in the church tower in Waingapu (Chapter 42) and showed three main clades within the conventional *Tyto alba*, corresponding to the European–African, North American and Australasian–South-East Asian populations (the last was represented by *delicatula* of Australia and the south-west Pacific, *javanica* of the Malay Peninsula to Wallace's line with *sumbaensis* on Sumba Island, and *delicatula* and other subspecies elsewhere). Christidis and Boles (2008) noted that Barn Owls from southern Asia through Australia and the Pacific differed from other Barn Owls and used the species name *javanica* Gmelin, 1788, which has priority over *delicatula* Gould, 1837. They adopted the English name Eastern Barn Owl for the present, although, they argued, alternatives should be considered. In contrast, König and Weick (2008) and Wink *et al.* (2008) recognised more than one species in the Barn Owl complex with *T. delicatula delicatula* in

Figure A.1 Possible breeding range of the Barn Owl in Australia.

Australia and its offshore islands, *contra* their earlier analysis, and placed *javanica* back with *alba*.

Range: Some ornithologists argue that Barn Owls are the most widely distributed land birds in the world. However, if the species is split into three or more species, this needs revision. The Barn Owl group occurs on all continents except Antarctica, in temperate and warm climates, but are absent from cold climates, so they do not occur in Greenland, the northern reaches of North America and Europe or most of Asia. They occur throughout Australia, though rarely in Tasmania. Possible breeding range (Figure A.1) needs revision.

Migration: In many parts of the world adult Barn Owls are sedentary and juveniles disperse, 1625 km in one instance. In Australia the true nature of their movements is poorly known, but they do not seem to have regular seasonal movements, as in a migration. More than other Australian owls, they appear to irrupt out of certain areas when prey numbers collapse. Barn Owl numbers increase in Canberra in some years, then they disappear for several years. We have yet to find them breeding in the ACT, but this may change.

Habitat, habits and food: Barn Owls prefer open areas. You often see them flying over open grassland or farmland, where they hunt mainly rodents. They are said to hunt mainly by ear, and the two earflaps or operculi are placed asymmetrically, at different levels on each side of the head, so owls can triangulate the location of rodents and catch prey in total darkness (Bunn *et al.* 1982). Eastern Barn Owls prey mainly on small terrestrial mammals, primarily small rodents, 61–99 per cent by number in a summary by Higgins (1999), and in inland Australia they also take dasyurid marsupials. For example, Debus *et al.* (2004) found that Barn Owls near Lake Frome in South Australia took 67 per cent mammals, including 4 per cent Fat-tailed Dunnart *Sminthopsis crassicaudata* and 62 per cent House Mouse, 3 per cent birds and 30 per cent lizards. Barn Owls also take bats, including fruit bats, and sometimes rabbit kittens, lizards, small birds and large insects. In some studies the House Mouse was the most common introduced prey taken in certain parts of Australia, but insignificant in other places where native Muridae (rats and mice) dominated. Barn Owls search in low, slow quartering flight, hovering before dropping onto prey, or by short-stay perch-hunting followed by a gliding attack to seize prey (Higgins 1999; Debus 2009).

Breeding: In Australia breeding is poorly known (Higgins 1999). Apparently they can breed at any time of the year if prey is abundant, in some cases twice in a year. Most use a tree hollow, or sometimes caves or empty mine shafts. Often they nest on a pile of old castings, feathers or faecal material. Unlike Barn Owls in Britain and in Indonesia (Olsen *et al.* 2009), they seldom nest in buildings in Australia. The reason for this difference is worth investigating.

Eggs: Three to six but occasionally as many as 12 eggs are said to be laid at two-day intervals. Incubation is about 30–34 days.

Young: They are brooded for about two weeks, and make short flights and climb after about 45 days (Higgins 1999).

Voice and calls: Higgins (1999) relied partially on Bunn *et al.* (1982) who recognised 15 calls, but this needs revision if the two owls are separate species. The commonest call is a long hissing screech given in one long note. They also give a rapid high-pitched twitter, said to be given by the male before mating, a food-offering call *e-e-e-e*, and a hiss when threatened. Young give a similar call when threatened. Courting Barn Owls give a variety of chirruping or wheezing calls.

Status: The Eastern Barn Owl is not nationally or globally threatened, being listed in Australia as Lower Risk – Least Concern (Debus 2009). They are common in suitable habitat and they are the common owl seen on islands north of Australia, like Sumba in Indonesia, or Solomon Islands. Numbers vary with prey abundance so they can be common or absent in an area in subsequent years. Barrett *et al.* (2007) recorded a 61 per cent decline in reporting rate in New South Wales between the first and second atlas of Australian birds.

Myths and questions: The taxonomy of this species is unclear. In Australia it is given a different common name and scientific name by Christidis and Boles (2008) versus Wink *et al.* (2008). Some subspecies may be full species, and those on small islands require special consideration. Also, it is not clear why they routinely nest inside buildings around the world, even on small islands like Sumba, but not in Australia.

Eastern Grass Owl *Tyto longimembris*

The species name *longimembris* is for its long legs.

Field recognition: The Eastern Grass Owl is a medium-sized owl with a very short tail and long legs. Length 30–38 cm; wingspan 100–115 cm; weight 310 g. They are superficially like Barn Owls, and like Barn Owls they can be found in open country, but these owls are most often seen flying, or perched on the ground, whereas Barn Owls often perch in trees and stand on fence posts in the open. Unlike Barn Owls, they have longer legs so their feet extend out past the end of their tails when they fly. They have longer wings than Barn Owls, apparently adapted to coursing over open ground as harriers do, instead of hunting from perches as Barn Owls do. They are the only Australian owl that routinely nests on the ground in open country. Though whitish below, they are darker above than a Barn Owl, charcoal or dark tawny, heavily blotched with lighter orange. Primaries and tail feathers are broadly barred with charcoal and orange (Hollands 1991; Higgins 1999; Debus 2009).

The **adult** has relatively small eyes for an owl set in a heart-shaped facial disc outlined with a pale, narrow ruff. The upperparts are dark brown with buff patches and fine pale spots. The underparts are white to buff-brown, with dark spotting. The face is correspondingly white to chestnut, with a small brown patch in front of each eye. The bill and cere are pinkish-white, and the lower legs and feet are dull pink to brownish-pink. The long tarsi are bare. The **juvenile** has buff down on the head and underparts, which is soon shed to reveal slightly darker plumage than that of the adult (Higgins 1999; Debus 2009).

Subspecies: König *et al.* (1999) and Wink *et al.* (2008) separated Grass Owls into two species, the Eastern Grass Owl and African Grass Owl. They recognised four subspecies of Eastern Grass Owl, including *T. longimembris longimembris* for Australian birds. Higgins (1999), following Schodde and Mason, considered African and Eastern Grass Owls to be the same species. Higgins recognised three subspecies: *capensis*; the nominate *longimembris* from north-eastern India, south-eastern China, South-East Asia, Indonesia, Philippines, New Caledonia, Fiji and northern and eastern Australia; and the poorly known *papuensis* in New Guinea. Christidis and Boles (2008) cited Rasmussen and Anderton (2005) who said there were numerous differences between *T. capensis* and *T. longimembris* (including plumage characters, proportions, egg size and at least some vocalisations) that

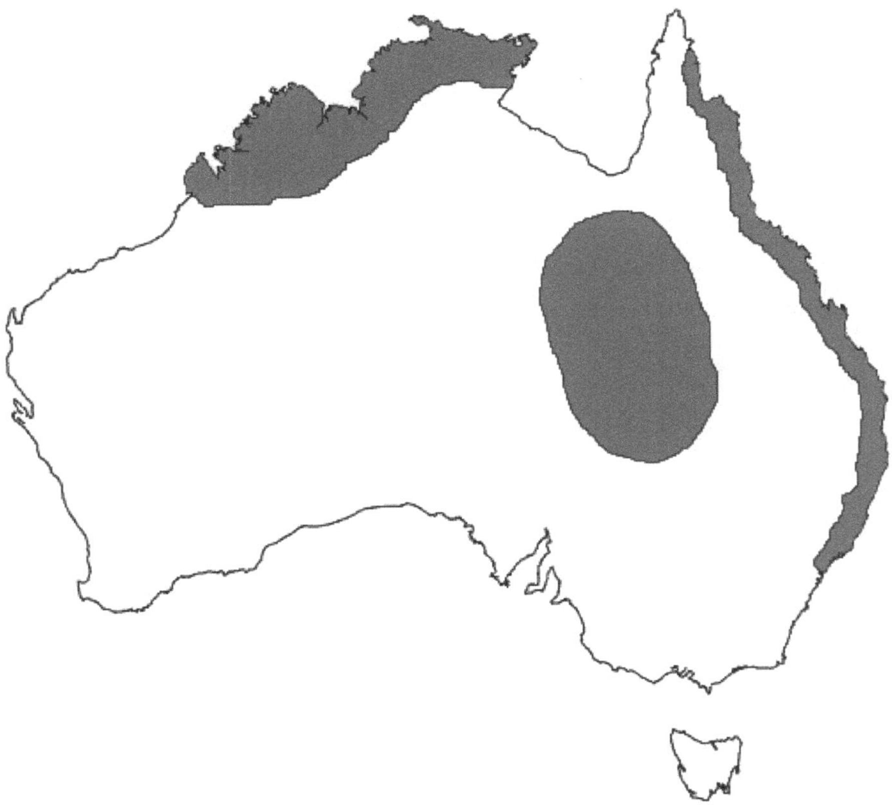

Figure A.2 Possible breeding range of the Grass Owl in Australia.

clearly indicated these were specifically distinct. Consequently, they treated Australian Grass Owls as *Tyto longimembris*, with the common name Eastern Grass Owl, and the DNA work of Wink *et al.* (2009) supported this.

Range: The range of the Eastern Grass Owl is poorly understood because the owl is so difficult to detect and study. The range seems to form an arc around the Bay of Bengal, India to China and South-East Asia, down the east Australian coast, but only in patches. Its Australian range is also poorly known. Higgins (1999) showed the owls in isolated patches in the far north around Darwin, Cape York Peninsula, down the Queensland and New South Wales coast to the ACT border and in two isolated patches in western New South Wales and north-east South Australia. The range in Simpson and Day (1999) and Hollands (1991) is more optimistic, showing a larger range in central Australia, and, unlike Higgins (1999), they include a large part of inland and Western Australia. Pizzey and Knight (2007) do not show them in western New South Wales. The map in König *et al.* (1999) covers more of northern Australia, but does not go as far south as the ACT (Figure A.2). Debus (2009) suggested that its core range appeared to be eastern

Queensland and northern New South Wales, and vagrants can occur almost anywhere, as with Letter-winged Kites.

Migration: Grass Owls seem to be nomadic, but it is not clear whether the various isolated populations mix. Breeding adults in core coastal areas appear to be resident, while juveniles are highly dispersive. Queensland dispersing birds establish temporary populations on the northern Australian coast, and in the eastern interior of Australia from the Barkly Tableland and Lake Eyre Basin to the inland plains of New South Wales in wet years, sometimes breeding there. Occasional vagrants reach south-western Australia and southern Victoria (Higgins 1999; Debus 2009). In short, we have little clear data on timing of breeding and more study is needed.

Habitat, habits and food: Grass Owls live in open, tussocky grasslands usually in treeless tropical and semi-tropical regions, on dry and wet ground. In Australia they often inhabit Bladey Grass *Imperata cylindrica* or sedges, often in swamp depressions, sometimes around cane fields. They have started to use young pine plantations. They roost and nest on the ground in rank vegetation associated with areas of dense tussock-grass, heath or reeds up to 2 m tall and seasonally flooded, in which they can hide in runways invisible from above. In coastal areas potential habitat may be indicated by Sword Grass *Gahnia sieberiana* or Common Reed *Phragmites australis*; in inland areas potential habitat may be indicated by Mitchell Grass (*Astrebla lappacea*), Cumbungi (*Typha* spp.) or Lignum (*Muehlenbeckia florulenta*) (Hollands 1991, 2008; Higgins 1999).

Grass Owls prey mainly on small terrestrial mammals, especially Muridae such as the Long-haired Rat *Rattus villosissimus*, Canefield Rat *Rattus sordidus* and Pale Field Rat *Rattus tunneyi* but also small dasyurid marsupials, occasionally small gliders and some birds and insects, and on the east coast, introduced House Mice and Black Rats (Higgins 1999).

They hunt by quartering over the ground, much like a harrier, and make quick plunges into the grass, in contrast to Barn Owls which sometimes hunt from perches. Prey is probably hidden and found when the owl is directly above it, or when the owls hear it (Hollands 1991).

Breeding: Little is known. Grass Owls probably can breed at any time during the year when food and conditions are favourable. They are said to form monogamous pairs in small territories, solitarily or in loose congregations with as little as 50 m between neighbours, and up to 40 pairs in 40 ha. In coastal northern Australia they often lay in March to June. They nest on the ground in dense tussocks of grass or sedges, and the nest, a scrape or pad under a tussock or shrub, is approached by a series of tunnels the owls make by pushing through the grasses on foot.

The owls lay eggs on the ground in tall dense grassland, sometimes near or surrounded by water.

Eggs: The clutch is three to eight eggs, more usually six to eight when food is abundant. Incubation takes about 31 days.

Young: Owlets are able to run, and leave the nest at 4–5 weeks old; they are fully fledged and able to fly when 2 months old. The period of dependence after fledging is not known, but it is likely to be some weeks if not months (Schodde and Mason 1980; Hollands 1991; Higgins 1999; Debus 2009).

Voice and calls: Hollands (1991, 2008) thought Grass Owls were more silent than Barn Owls. They make a hissing scream, in loudness and harshness midway between that of a Barn Owl and Masked Owl, a high cricket-like chirruping uttered by the male when returning to the nest, the bill-clapping common to many owls in aggression and a number of other calls. *Tyto* owls appear to achieve this bill-clapping by catching the lower mandible under the ridge inside the hook of the upper mandible, then forcefully snapping the bill shut with a loud clap (Stephen Debus personal communication).

Status: Eastern Grass Owls are seldom seen tropical owls. Consequently, their numbers and status are difficult to determine. They are not nationally or globally threatened, the assumed subspecies *longimembris* being listed in Australia as Lower Risk - Least Concern (Garnett *et al.* 2011). However, it is listed as Vulnerable in New South Wales and South Australia and Rare in Western Australia. It may be threatened by destruction of coastal habitat, overgrazing of inland habitat, fire and intensive cropping. Grass Owls and their eggs and young are vulnerable to predation by foxes and feral cats, and also to accidental destruction by agricultural machinery and livestock. Some are killed by collisions with vehicles and barbed-wire fences. There might be 5000 breeding pairs in Australia (Debus 2009).

Myths and questions: König *et al.* (1999), Wink *et al.* (2009), Higgins (1999), and Christidis and Boles (2008) disagree about Grass Owl taxonomy, with Higgins calling the African Grass Owl and Eastern Grass Owl one species. Further DNA analysis is needed.

Higgins (1999) and König *et al.* (1999) said that Grass Owls did not occur on Sumba, but Jones *et al.* (1994) saw them, as we did in 2002 (Chapter 42). There is still no museum specimen from Sumba, so the taxonomy and status of these Grass Owls are unknown.

Masked Owl *Tyto novaehollandiae*

The species name *novaehollandiae* is for New Holland (New South Wales).

Field recognition: The Masked Owl is a medium-sized to large owl, thick-set and larger than a Barn Owl or Grass Owl, closer to the size of a Sooty Owl in dimensions, and decreasing in size from Tasmania to the Tropics. In plumage it is the most variable Australian *Tyto*. Males are about 33–41 cm long with a wingspan of 90–110 cm, females 39–50 cm long with a wingspan of 96–128 cm; in Tasmania males average about 633 g and females 841 g; for mainland Australia males average about 476 g and females 630 g (Higgins (1999). The **adult** has large eyes set in a rather rounded, heavily rimmed facial disc. The colouring inside the mask can be white to tawny or chestnut. The feet are massive and powerful, though less so than in the Sooty Owl. Unlike Barn or Grass Owls, they have fully feathered tarsi, extending down to the toes (Higgins 1999; Debus 2009).

The **adult** is a large *Tyto* with three main colour morphs – light, intermediate and dark, connected by a range of intermediate forms (Higgins 1999). The darkest form has upperparts dark-slate or blue-black; underparts tawny and chestnut. The facial disc is tawny, heavily edged with black and rounder than the edging of the facial disc on a Barn Owl. The lightest form has upperparts patterned with a mixture of slate, fawn or pale gold and white. Some birds are similar to Barn Owls, but with a barred or reticulated pattern on the upperparts always present in the Masked Owl and absent in the Barn Owl. This heavy barring extends to flight and tail feathers in the Masked Owl. Sides of the head are white, facial disc is white and underparts are white, finely speckled with black. Intermediate forms fall between these two and there is considerable variation and gradation between the lightest and darkest forms. Females are significantly larger and heavier than males (Hollands 1991; Higgins 1999). In flight the Masked Owl shows a large head, long broad wings and short tail with the toes just protruding beyond the tip. The **juvenile** has cream down on the head and underparts, which is soon shed to reveal slightly darker plumage compared with adults of the equivalent colour form (Higgins 1999; Debus 2009).

Subspecies: Higgins (1999) recognised eight subspecies, four of which occur in Australia – *novaehollandiae* in south and eastern Australia; *castanops* in Tasmania, and introduced to Lord Howe Island; *kimberli* in northern Australia; and *melvillensis* on Melville Island and possibly Bathurst Island. In contrast, König and Weick (2008) did not recognise *melvillensis*, but recognised *kimberli* in northern Australia and *calabyi* in New Guinea and argued that *castanops* in Tasmania is a separate species, partly because females are much larger and always darker than males and because of the cytochrome b analysis of Wink *et al.* (2008). Christidis and Boles (2008), however, did not recognise any separate species within the complex.

Range: Because Masked Owls are secretive and nocturnal their distribution is imperfectly known. In Australia they range from Tasmania north into southern Victoria, along the east coast to Cape York and west to north-west Western Australia. There is an isolated population in south-west Western Australia (Hollands 1991; Higgins 1999). König *et al.* (1999) give a slightly different distribution, with two isolated populations in Western Australia, one inland and one in the coastal south-east.

On the mainland it is said to be most abundant in coastal regions (Figure A.3). It also occurs in southern New Guinea. The subspecies *novaehollandiae* occurs in

Figure A.3 Possible breeding range of the Masked Owl in Australia.

eastern and southern Australia, *kimberli* in northern Australia, *castanops* in Tasmania, and *melvillensis* on Bathurst and Melville Islands. The forms in the Moluccas, Tanimbar Islands and Admiralty Islands are sometimes regarded as full species. Masked Owls were introduced to Lord Howe Island in a failed attempt to control rats and they have established there. Efforts are under way to eradicate them (Higgins 1999; Debus 2009; Milledge *et al.* 2010).

Migration: Said to be sedentary (Higgins 1999).

Habitat, habits and food: The Masked Owl is said to live in a diverse range of sclerophyll forest and woodland that provide tall or dense mature trees with suitable hollows, often near ecotones with open areas such as grassland or heath, and typically with a grassy ground layer or a mosaic of sparse and dense groundcover. It is sometimes seen in pairs or family groups. Roosting sites are tree hollows, caves or dense foliage and, rarely, buildings (Higgins 1999; Debus 2009).

Masked Owls feed mostly on small to medium-sized terrestrial mammals, including rodents, but also arboreal or climbing mammals (Higgins 1999). In south-eastern New South Wales they took 4.6 per cent arboreal mammals, mainly Sugar Gliders and Common Ringtail Possum, around 89.4 per cent terrestrial mammals, mainly Bush Rat, Brown Antechinus *Antechinus stuartii* and Dusky

Antechinus *Antechinus swainsonii*, and 3.8 per cent birds. On the central New South Wales coast they took 79.5 per cent terrestrial mammals and 20.5 per cent birds (Kavanagh 2002).

Breeding: Like other *Tyto* owls, they are said to breed opportunistically when food is available (Hollands 1991, 2008), but in Tasmania they are highly seasonal (Mooney 1997). In south-east Australia they tend to lay in early autumn to late winter, with some laying in spring. When food is scarce, they may not breed for that year. They use large tree hollows, and are said to nest on ledges in caves (Higgins 1999).

Eggs: One to four eggs, usually two or three, are laid on sand, soil or debris. Incubation is said to start on the first egg laid and to last 33 to 35 days (Higgins 1999).

Young: Schodde and Mason (1980), and Hollands (1991, 2008) said young fledged at 10 to 12 weeks, but Debus (2009) thought this was too long and recorded 8 to 9 weeks for captives. However, fledging and branching are not separated in these accounts, and actual fledging may be partly determined by the nest site.

Voice and calls: The Masked Owl's characteristic call is a loud, harsh screech, similar to a Barn Owl but louder, deeper and harsher, lasting about 2 seconds. Other calls include cackling, chattering and rattling sounds. The juvenile begging call is a repeated, brief hissing screech (Hollands 1991, 2008; Higgins 1999).

Status: The Masked Owl is not globally threatened, being listed in Australia as Lower Risk – Least Concern (Garnett *et al.* 2011). However, they considered the assumed subspecies *melvillensis* Endangered, *kimberli* as Near Threatened (listed Vulnerable in Queensland), *novaehollandiae* as Near Threatened (listed Vulnerable in New South Wales and South Australia, and Endangered in Victoria), and *castanops* Endangered and now listed in state legislation (Debus 2009).

Myths and questions: König *et al.* (1999), König and Weick (2008) and Wink *et al.* (2008, 2009) argued that Tasmanian Masked Owls are a separate species from mainland Masked Owls. This question needs further DNA analysis (Christidis and Boles 2008). Maps for the distribution of the species differ widely between König *et al.* (1999), del Hoyo *et al.* (1999), Higgins (1999), Garnett and Crowley (2000) and Kavanagh (2002). The precise range seems to be unknown.

Sooty Owl *Tyto tenebricosa*

The species name *tenebricosa* is from the Latin *tenibricosus* for dark or gloomy.

Field recognition: The Sooty Owl is a medium-sized to large owl, depending on sex, with very large, dark eyes set in a round facial disc. Males are 33–36 cm long with a wingspan of 82–100 cm, females 38–43 cm long with a wingspan of 101–118 cm. Recorded weights, mostly from museum specimens, average 490 g for males and 890 g for females (Higgins 1999), but healthy live birds are heavier, a male 650 g, and females 990–1250 g (Debus 2009). The **adult** is sooty-grey above, finely speckled and spotted with white and with a rounded to heart-shaped facial disc thickly outlined in black and suffused with grey. The eyes are black-brown. Underparts are sooty-black to dark grey finely spotted with white. The rather large head and short tail give the owl a top-heavy appearance. The bill is cream, sometimes with dusky sides, the cere is cream, sometimes with a pink tinge, and the massive feet are grey-brown. In flight its large head, broad rounded wings and short tail with the toes just protruding beyond the tip are diagnostic.

The **juvenile** has grey down on the head and underparts, which it sheds to reveal slightly darker plumage than that of the adult (Hollands 1991, 2008; Higgins 1999).

Subspecies: Though a number of writers considered the 'Lesser Sooty Owl' a separate full species *Tyto multipunctata* (Schodde and Mason 1980; König *et al.* 1999; Higgins 1999; Hollands 1991, 2008), DNA evidence in Norman *et al.* (2002) indicated that *multipunctata* was a subspecies on the north Queensland coast. It is noteworthy that the sonograms of Lesser Sooty Owl and Sooty Owl in Higgins (1999) are the same, indicating they are one species. This treatment was followed by Christidis and Boles (2008) who recognised three recently diverged subspecies – *tenebricosa*, *multipunctata* and *arfaki*. The subspecies *multipunctata*, or 'Lesser Sooty Owl' is smaller, more lightly built, with proportionally smaller feet. Males are 31–35 cm long with a wingspan of 79 cm, females 35–38 cm long with a wingspan of 86–97 cm. Males average 370 g, females 560 g. The **adult** has large eyes set in a rather rounded facial disc. It is grey with a pale face and dark rings around the eyes. It is profusely and finely spotted, vermiculated and barred on the upperparts, and white densely marked with fine wavy bars on the breast, with a pale belly. The bill is cream to pale grey, sometimes with dusky sides, the cere is pink to pink-grey, and the feet are grey to grey-brown (Higgins 1999; Debus 2009). However, this taxonomic treatment is complicated by Wink *et al.* (2008) who used the cytochrome b gene to argue that Lesser Sooty Owls are a different species. Clearly this needs further analysis.

The third subspecies, *T. t. arfaki* in New Guinea, is generally smaller than southern Sooty Owls and said to be morphologically intermediate between the other subspecies. The legs have denser brown barring and the underparts have finer white spots, with a darker, sootier facial mask. Wing length is about 255 mm for males, 281 mm for females. Weights of two males were 422 g and 460 g (Higgins 1999). Wink *et al.* (2008) appear to retain this subspecies in the Sooty Owl species *tenebricosa*, as did Christidis and Boles (2008), but this needs confirmation.

Range: The Sooty Owl occurs in coastal and mountain regions of south-eastern Australia, from mid-eastern Queensland (Eungella National Park) to southern Victoria (Figure A.4). It also occurs throughout New Guinea, as *T. t. arfaki*; the race *T. t. multipunctata* occurs in coastal north-eastern Queensland between Cooktown and Townsville (Higgins 1999).

Figure A.4 Possible breeding range of the Sooty Owls in Australia.

Migration: Said to be sedentary.

Habitat, habits and food: The Sooty Owl is the most nocturnal of the Australian owls. It is often found in, but not restricted to, tall, old-growth montane forest of subtropical as well as temperate rainforest and in wet sclerophyll forest. It is sometimes seen in pairs or family groups. It is associated with moist, sheltered gullies and tree species such as Mountain Ash *Eucalyptus regnans*, or Mountain Grey Gum *E. cypellocarpa* (Higgins 1999; Debus 2009). Roost and nest sites are often located in gullies, tree hollows, caves, cliff-ledges or crevices, rock overhangs, or dense vegetation, tree-ferns and gloomy, vine-tangled gullies (Higgins 1999; Debus 2009).

The Sooty Owl preys mainly on arboreal mammals, mostly marsupials, including small possums and gliders, as well as terrestrial mammals, primarily rodents and small dasyurid marsupials, but also bandicoots, potoroos and rabbits. It sometimes takes birds, lizards and large insects.

Kavanagh (2002) found that Sooty Owls in New South Wales took roughly, by number, 45 per cent arboreal mammals, 50 per cent terrestrial mammals, the remaining 5 per cent mostly birds.

Breeding: There is little information on Sooty Owl breeding, and no detailed quantitative studies. They appear to form monogamous pairs during the breeding season that are resident in large home-ranges of at least 350 to 1000 ha or more, depending on quality of habitat and abundance of prey. The owl's laying season varies and is perhaps opportunistic. They are said to lay usually from January to June with peaks in April, but there is some laying in August–September. They have one brood per year in the wild (Higgins 1999; Debus 2009). Pairs nest in a large hollow in a big old forest tree, particularly a smooth-barked eucalypt like *Eucalyptus grandis, E. cypellocarpa, E. bicostata, E. viminalis,* or sometimes a fig, in a moist gully. The entrance has been measured as 10–50 m above the ground, with the chamber more than 30 cm wide and 0.5 to 5 m deep. The owl also nests on ledges in caves (Higgins 1999; Hollands 2008; Debus 2009).

Eggs: The clutch is one or two eggs. The incubation period is said to be 42 days, although it could be shorter as the similarly sized Masked Owl is about 35 days (Higgins 1999; Hollands 2008; Debus 2009).

Young: The nestling period is said to be about 3 months (Fleay 1968; König *et al.* 1999; Higgins 1999; Debus 2009) but 90 days is the longest nestling period recorded for any owl species, even longer than the much larger Snowy Owl *Nyctea scandiaca* or Eurasian Eagle Owl *Bubo bubo* (König *et al.* 1999). The nestling period is probably closer to 8 or 9 weeks as with Masked Owls. The young are said to be dependent on their parents for 3–5 months after fledging. In New South Wales, seven young were reared from eight nesting attempts, with an average brood size of 1.2 fledglings per successful nest (Higgins 1999; Debus 2009).

Voice and calls: The Sooty Owl's characteristic call is a piercing, downscale shriek lasting about 1.5 seconds, a long, descending siren-like scream or wail. Fleay likened it to the sound of a falling bomb without the explosion. The Sooty Owl also gives a harsh screech, like that of the Masked Owl, and a long series of grating trills, clicks and rattles. The juvenile begging call is a long, descending, wheezy screech (Fleay 1968; Hollands 1991; Higgins 1999).

Status: The Sooty Owl is not globally threatened, being listed in Australia as Lower Risk – Least Concern (Garnett *et al.* 2011). However, subspecies *tenebricosa* is listed under state legislation as Rare in Queensland, and Vulnerable in New South Wales and Victoria. The species is uncommon, though not as rare as formerly believed, and is widespread in south-eastern Australia. It is threatened by clearance and fragmentation of habitat, by intensive forestry practices that remove hollow trees over wide areas, and by fire. It persists in native wood-production forests by breeding in unlogged areas such as rainforest or gully reserves and foraging in selectively logged or regenerating areas. There might be 5000 breeding pairs in Australia (Debus 2009).

Myths and questions: Because pairs are so difficult to locate and study, there are still questions about the range and abundance of this species. Some texts still consider the Lesser Sooty Owl a separate species. Three months for the nestling stage is either unique, or incorrect.

Ninox owls

The genus name *Ninox* is combined from *Nisus*, sparrowhawk, and *Noctua*, owl. Australia, including Christmas Island, has five or six *Ninox* owls, depending on taxonomic treatment. Unlike some *Tyto* species, all prefer forest or woodland to open grassland.

Rufous Owl *Ninox rufa*

The specific name *rufa* is Latin for red or ruddy.

Field recognition: The Rufous Owl is a large owl, about 46–57 cm long with a wingspan of about 115 cm. In eastern Queensland males are said to average about 950 g and females 660 g; in northern Australia males average 1450 g and females 950 g, with big males said to be up to 1900 g (Debus 2009). It has a relatively small head and long tail, though the head can look larger when the head-feathers are raised in alarm. The **adult** is rufous-brown with fine bars on the upperparts, and cream with finely barred rufous on the underparts. It has yellow eyes in a prominent black-brown mask, and fully feathered legs with large yellow feet. The bill is pale blue-grey, and the cere is pale grey. The **juvenile** is white and downy on the head and underparts, with dark eye-patches, but when fully feathered is similar to adults, though upperparts may be slightly warmer. The bill is grey, and the cere is greyish-green (Higgins 1999; Debus 2009).

Subspecies: König *et al.* (1999) recognised six subspecies, but Higgins (1999) recognised five – *rufa*, the nominate form from north-west Western Australia and the Northern Territory, *meesi* from the Cape York Peninsula, *queenslandica* from coastal north-east Queensland, *humeralis* from New Guinea and Waigeo Island, and *aruensis* from the Aru Islands in Indonesia. Christidis and Boles (2008) did not discuss *N. rufa* subspecies.

Range: The Rufous Owl occurs in coastal northern Australia from the Kimberley in Western Australia and Top End of the Northern Territory to the mid-eastern Queensland coast, extending at least as far south as Rockhampton and the Blackdown Tablelands (Figure A.5). It is said not to occur on islands in the Torres Strait. The southern limit is uncertain, owing to the similarity of its calls to those of the Powerful Owl. It also occurs in New Guinea and on some of its satellite islands (Higgins 1999; König *et al.* 1999; Debus 2009).

Migration: None known.

Habitat, habits and food: The Rufous Owl lives and hunts in tropical rainforest, and sclerophyll forest and woodland near rainforest in lowlands up to at least 1200 m. Its roost-sites are located in sheltered groves of dense, shady trees.

Figure A.5 Possible breeding range of the Rufous Owl in Australia.

Rainforest trees, paperbarks (*Melaleuca*) or gallery rainforest on watercourses through eucalypt forest or woodland may indicate potential roosting areas. The owls also occur in swamps, mangroves and large botanic gardens in cities, e.g. Darwin and Townsville, and at night they range into suburban gardens (Higgins 1999; Debus 2009).

Resident pairs of Rufous Owls have traditional day roost trees, which can be close to or distant from the nest depending on prey, relationships with neighbouring Rufous Owls and the time of year. By day they are seen in pairs, family groups or alone. If flushed, members of the pair re-establish contact with soft hooting calls. Adults can defend their nest against humans, sometimes defending fledglings that are roosting near the ground. During the day, owls often hold part of their prey, such as a possum, flying-fox or large bird, and sometimes eat it in daylight (Hollands 1991; Higgins 1999; Debus 2009).

In courtship members of a pair sidle up to each other and allopreen, call with trilling or soft hooting calls, and the male feeds the female, which takes the food. A long series of soft, monotonous hooting notes, or bleating calls, precede mating, which often takes place on a branch near the nest with quavering squealing notes (Debus 2009).

In response to a persistently intruding owl, as simulated by recorded playback, a defending Rufous Owl often replies with *woo-hoo* calls, and may approach in a threatening manner (Higgins 1999; Debus 2009).

The Rufous Owl preys mainly on medium-sized arboreal and terrestrial mammals such as possums and gliders and on flying-foxes where these are common. It also takes a range of birds including kookaburras, scrubfowl and frogmouths, and some large insects. Its hunting behaviour is not well known, but is probably similar to that of the Powerful Owl. It takes prey in a direct flying attack or swoop to snatch it from trees, shrubs, from the air or the ground. Flying-foxes are sometimes caught aerially in a tail-chase, and it sometimes flushes and takes insects by crashing into foliage (Higgins 1999; Debus 2009). A study in Kakadu (Estbergs and Braithwaite 1985) showed they took, by number, about 3 per cent insects, 9 per cent birds, 30 per cent Fawn Antechinus *Antechinus bellus*, 5 per cent Sugar Gliders *Petaurus breviceps*, 6 per cent flying-foxes, and about 41 per cent Muridae (placental rats and mice) (Higgins 1999; Debus 2009).

Breeding: This is poorly known. They are said to lay in winter and spring, from June to October. The nest is a large hollow said to be up to 60 cm wide and 2 m deep in a big old tree, up to 40 m above the ground.

Eggs: The clutch is two eggs, rarely one. Incubation is by the female alone and said to take 36–38 days.

Young: The nestling period is said to be 6 to 8 weeks; Hollands (2008) gives 50 days. Broods of two owlets often fledge, although some brood reduction occurs. Causes of nesting failure include invasion of the nest hollow by green ants and predation on owlets by goannas (monitor lizards *Varanus* spp.). Young are dependent on their parents for several months, sometimes up to the start of the next breeding season (Schodde and Mason 1980; Hollands 1991; Higgins 1999; Debus 2009).

Voice and calls: The Rufous Owl's characteristic call is a slow double hoot *woo-hoo*. Each note is about half a second in duration, with a half-second interval between the notes in each couplet (Higgins 1999); the couplet is repeated at intervals of several seconds. Sometimes a single hooting note is given. Other calls include guttural bleating. The juvenile begging call is a cricket-like trill, more like that of the Barking Owl than the Powerful Owl. The Rufous Owl's hooting call is said to be somewhat more nasal and subdued than that of the Powerful Owl, lacking the ringing, far-carrying quality (Debus 2009). Nevertheless, calls of the two owls can be confused.

Status: The Rufous Owl is not globally threatened, being listed in Australia as Lower Risk – Least Concern. However, the assumed subspecies *meesi* on the Cape York Peninsula and *queenslandica* in Queensland are listed as Rare and

Vulnerable, respectively, under state legislation (Debus 2009), and as Least Concern in the latest national assessment (Garnett *et al.* 2011). The Rufous Owl is uncommon but widespread in the Tropics. It is threatened by clearance and fragmentation of habitat, by fire and possibly by intensive forestry practices that remove hollow trees over wide areas. There are said to be possibly 1000 breeding pairs of *meesi* (Garnett *et al.* 2011) and possibly 3000 breeding pairs of *queenslandica* (Garnett and Crowley 2000), though Debus (2009) suggests a possible total of 3000 pairs for Australia, the majority in Queensland.

Myths and questions: The differences in taxonomy presented by König *et al.* (1999) versus Higgins (1999) need further analysis. Christidis and Boles (2008) list no subspecies.

Powerful Owl *Ninox strenua*

The species name *strenua* is from the Latin *strenuus*, vigorous because of its very large size.

Field recognition: The Powerful Owl is a large owl, 50–60 cm long with a wingspan of 110–140 cm. Males are said to average 1450 g, females 1250 g (Higgins 1999), though big males can be up to 2200 g and are larger than most Australian diurnal raptors (Debus 2009). For an owl, it has a proportionally small, round head and a long tail. The **adult** has a hawk-like appearance; it is brown, narrowly barred on the upperparts and white, coarsely marked with V-shaped bars, on the underparts. It has prominent orange-yellow eyes and fully feathered legs with large yellowish feet. The bill is dark grey, paler at the base, and the cere is light blue-grey. The **juvenile** is white and downy on the head and on the underparts, with dark eye-patches and fine shaft-streaks on the breast (Higgins 1999; Debus 2009).

Subspecies: None recognised.

Range: The Powerful Owl occurs in south-eastern Australia from mid-eastern Queensland (Clarke Range north of Eungella) to western Victoria, and recently in south-east South Australia, and from the coast to the inland slopes of the Great Divide and, in some sufficiently well-vegetated areas, along rivers to the edge of the inland plains (Figure A.6) (Debus 2009).

Migration: None known.

Habitat, habits and food: The Powerful Owl lives in sclerophyll forest, including that with rainforest trees in the mid-storey, and dry sclerophyll forest or woodland, extending inland through riverine woodland. Its roost and nest sites are usually located in gullies, including the heads of minor drainage lines, in sheltered groves of mid-storey trees, or sometimes pine plantations and paperbark (*Melaleuca*) thickets on inland rivers (Debus 2009). At Kioloa on the south coast of New South Wales, Powerful Owls live in tall forest dominated by Spotted Gum *Eucalyptus maculata* and Blackbutt *E. pilularis*. In Victoria, they use box and ironbark forest, Mountain Grey Gum *E. cypellocarpa*, Mountain Ash *E. regnans*, Candlebark *E. rubida*, Messmate *E. obliqua* and River Red Gum *E. camaldulensis*.

Figure A.6 Possible breeding range of the Powerful Owl in Australia.

In the ACT we found them in forests dominated by Manna Gum *E. viminalis*, Brown Barrel *E. fastigata*, Narrow-leafed Peppermint, *E. radiata*, Broad-leaved Peppermint *E. dives* and Mountain Gum *E. dalrympleana*. For daytime roosts, they and Southern Boobooks often use Native Cherry *Exocarpos cupressiformis*, Blackwood *Acacia melanoxylon*, Red Turpentine *Syncarpia glomulifera*, Coachwood *Ceratopetalum apetalum*, Lilly Pilly *Acmena smithii*, Sassafras *Doryphora sassafras*, or groves of she-oaks (*Casuarina*) (Olsen and Rehwinkel 1995; Higgins 1999; Debus 2009).

Powerful Owls also live in urban bushland and large botanic gardens in cities such as Melbourne, Sydney and Canberra. At night they range into adjacent suburban gardens. They can be agile and buoyant, like a giant Southern Boobook, but they can appear clumsy when attacking small agile prey such as Sugar Gliders. They glide, showing long, broad, deeply fingered wings and a long, rounded tail (Higgins 1999; Debus 2009; Olsen *et al.* 2011).

Resident pairs of Powerful Owls have traditional roosting groves, near or away from their nesting gully depending on the time of year, prey and relationships with neighbouring Powerful Owls. To avoid discovery by enemies such as mobbing birds, they choose secluded day-roosts under dense leafy canopy. They roost singly, in pairs or in family groups and tend to roost in the top one-third of a tree or shrub. By day, an owl sometimes betrays its presence by hooting softly or growling at a human intruder. Usually it only stares, but adults sometimes defend fledglings. One member of the family, often the male, sometimes holds part of the previous night's prey such as a possum, glider or large bird like a parrot, kookaburra, currawong or crow (Fleay 1968; Higgins 1999; Debus 2009; Olsen *et al.* 2011).

At dusk, members of a pair sometimes call to each other and, if they are roosting together, they allopreen. If one is holding prey, they may share it. As dusk deepens, they might advertise their territory or maintain contact with bouts of *woo-hoo* calling. After dusk, or before dawn or, some claim, even during the day on hot days, they bathe in pools of water in creeks, though daytime bathing may be a feature in captive owls rather than wild owls, which would be mobbed by enemies. At dawn, they return to their roosts and sometimes advertise the site with calling in the growing light before sunrise.

In courtship, members of a pair sometimes sidle up to each other and allopreen; they may call with bleating or soft *woo-hoo* calls, and the male feeds the female. A long series of soft, monotonous hooting notes leads to mating, which takes place on a branch near the nest and culminates with a rabbit-like squealing call (Debus 2009).

In response to a persistently intruding owl, as simulated by call playback, a defending Powerful Owl may approach with heavy 'thumping' landings on branches and lowered wings, giving menacing *woo-hoo* calls (Debus 2009).

The Powerful Owl preys mainly on mammals, particularly medium-sized arboreal marsupials such as possums and gliders, but also on flying-foxes, other mammal species, bird species and insects, the latter sometimes hawked in flight. In south-eastern New South Wales Kavanagh (2002) found they took, by number, 48.6 per cent Common Ringtail Possums, 30 per cent Greater Gliders, 2.1 per cent Yellow-bellied Gliders, 9.1 per cent Sugar Gliders, 0.2 per cent Mountain Brushtail Possums, and 1.3 per cent Common Brushtail Possums. In contrast, Pavey *et al.* (1994) found near Brisbane, Queensland, that Powerful Owls took these marsupials in smaller percentages, but took about 67 per cent bats – Black *Pteropus alecto*, Grey-headed *P. poliocephalus* and Little Red *P. scapulatus* Flying-foxes. They search for prey by perching in trees, watching and listening and sometimes flying to a new perch. They take prey in a direct flying attack or swoop to snatch it from trees, out of the air, from shrubs or, rarely, from the ground (Higgins 1999; Debus 2009; Olsen *et al.* 2011).

Breeding: Powerful Owls are said to form monogamous, lifelong pairs (Higgins 1999), and there is some evidence for this (Hogan and Cooke 2010), but this needs verification with long-term observation (decades) of several marked pairs and specific territories. More likely, certain females change territories more often than males do, as with Southern Boobooks, perhaps after being paired for many years, and only certain pairs mate for life. They are resident in large home ranges of about 600 ha in highly productive, moist coastal forest to 4774 ha in dry inland forest (Soderquist and Gibbons 2007). They lay eggs in autumn and winter, usually in May–June, though slightly earlier in the north of their range. Pairs breed solitarily. The nest is a large hollow up to 1 m wide and 2 m deep in a big old tree, usually live, but sometimes dead, up to 45 m above the ground.

Eggs: The clutch is two eggs, rarely one. Incubation takes 35–40 days, and the nestling period is said to be 8–9 weeks (Fleay 1968; Pavey *et al.* 1994; Higgins 1999; Debus 2009).

Young: Breeding success has been measured as: 94 per cent of attempts, with 1.4 young fledged per pair per year in southern Victoria (McNabb 1996); 0.4–0.5 fledglings per pair per year in inland Victoria (Hollands 1991); 0.8 young per attempt in New South Wales; and 76 per cent of attempts in coastal New South Wales with half of such broods having two young (Debus and Chafer 1994). Seven young fledged in five years by one pair in Queensland (Pavey *et al.* 1994).

Young are dependent on their parents for several months after fledging, sometimes up to the start of the next breeding season. One female bred successfully in the wild at one year old (Higgins 1999; Debus 2009; McNabb *et al.* 2007).

Voice and calls: The Powerful Owl's characteristic call is a deep, slow double hoot *woo-hoo*, with a far-carrying quality. Each note is about 1 second in duration, with a half-second interval between the notes in each couplet; the couplet is repeated at intervals of several seconds (Higgins 1999). In males, it is said that the second note is slightly lower than the first, and in females slightly higher (Higgins 1999), though this needs verification. This could be single-blind tested in the field, as in Olsen *et al.* (2002a), to determine if calls are a reliable tool for sexing owls. Sometimes a single hooting note is given. Other calls include guttural bleatings and rumblings. The juvenile begging call is a falsetto trill, incongruous in so large a bird. The Powerful Owl's hooting call is similar to that of the Rufous Owl, so in the zone of overlap in eastern Queensland calling birds should be sighted before identification is confirmed (Debus 2009).

Status: The endemic Powerful Owl is not nationally threatened, being considered Lower Risk – Least Concern (Garnett *et al.* 2011). It is uncommon, though not as rare as formerly believed, and widespread in south-eastern Australia. However, it is listed under state legislation as Vulnerable in Queensland and New South Wales, and Endangered in Victoria. It may be threatened by clearance and fragmentation of habitat, by intensive forestry practices that remove hollow trees over wide areas, and by fire. The species persists in native wood-production forests. There might be 3500 breeding pairs Australia-wide, but there is a need to monitor their numbers and breeding success (Debus 2009).

Myths and questions: Claims that they always mate for life need confirmation. Consistent differences between male and female territorial calls need to be verified.

Barking Owl *Ninox connivens*

The species name *connivens* is from the Latin for winking or blinking.

Field recognition: The Barking Owl is a medium-sized hawk-like owl, 39–44 cm long with a wingspan of 85–120 cm, and varying in size, smaller in the north than in the south. It is intermediate in size and build between the Powerful Owl and Southern Boobook, giving more the impression of a small Powerful Owl. In eastern Australia males are said to weigh about 695 g and females about 592 g (Higgins 1999), although these weights are mostly from museum specimens and are likely to be suspect. The **adult** is brown, spotted white on the wings, with barring in the wings and tail, and white with coarsely streaked brown on the underparts. It has prominent yellow eyes and fully feathered legs with large yellowish feet. The bill is pale grey, darker at the tip, or (less commonly) straw yellow, and the cere is pale grey to greyish-yellow or straw-yellow. The **juvenile** has 'messy' dark streaking on the underside, and a downy collar when newly fledged (Higgins 1999; Debus 2009).

Subspecies: König *et al.* (1999) recognised five subspecies, but Higgins (1999) recognised four subspecies that varied in size and tones of plumage, but not in pattern of plumage. Following Mees (1964), Higgins' arrangement for Australia was: the nominate *connivens* subspecies in south-west and eastern Australia, south of the base of the Cape York Peninsula; *peninsularis* on the Cape York Peninsula and the islands of the Torres Straight; *occidentalis* in the north-west. König *et al.* (1999) recognise *rufostrigata* for the Moluccas, and *assimilis* for parts of New Guinea. Christidis and Boles (2008) mention no subspecies.

Range: The Barking Owl is said to occur in northern and eastern Australia from the Pilbara and Kimberley of Western Australia, across the Top End of the Northern Territory, and south through extreme eastern South Australia in river gums on permanent water and box woodland, to southern Victoria, with an isolated population in south-western Australia (Figure A.7), and populations in New Guinea and the Moluccas (Higgins 1999; Debus 2009).

Migration: None known.

Habitat, habits and food: Barking Owls live in dry sclerophyll forest and woodlands of tropical, temperate and semi-arid zones often dominated by *Eucalyptus* (Higgins 1999), and swamps dominated by paperbarks (*Melaleuca*), as well as dry sclerophyll forest and woodland typically dominated by red gums or box trees. It is said to be an edge species, associated with wetlands or watercourses, open areas and old eucalypts with large hollows, though it also occurs in the interior of extensive dry inland forest, e.g. the Pilliga in New South Wales. It roosts in gallery forest of *Casuarina* or rainforest trees, in River Red Gum forest, in other

Figure A.7 Possible breeding range of the Barking Owl in Australia.

eucalypts such as boxes with dense crowns, or in streamside thickets of *Acacia*. Its flight is quick, agile and direct, with some gliding, showing a slender build with moderately long wings and tail (Debus 2009).

Resident pairs of Barking Owls have roost trees near or away from their nest depending on the time of year, weather, stage of the breeding cycle, prey or relationships with neighbouring Barking Owls. At its roost the owl might betray its presence by growling at human intruders. The adults are likely to vigorously defend their young against humans. Sometimes an owl holds partly eaten prey at the roost. Barking Owls are sometimes active during the day, particularly in the late afternoon or in overcast conditions, when they 'bark' from their roost or hunt opportunistically near the roost, eating their catch in daylight (Debus 2009).

At dusk, members of a pair are said to sometimes call to each other with soft growling calls and duets of soft barking. As dusk deepens, they might advertise their territory, duel with neighbours or maintain contact with persistent bouts of louder 'barking'. At dawn, they return to their roosts and sometimes call in the growing light before sunrise, after the dawn chorus of diurnal birds has started (Debus 2009).

The Barking Owl preys on mammals, birds and large insects. They take rabbits where these are common and there are few alternative prey, but they seem to prefer arboreal marsupials such as small possums and gliders, or native rats in the 'channel country' of the eastern arid zone. They also take many birds up to the size of kookaburra, magpie or Tawny Frogmouth, and are partly insectivorous. For example, near Armidale New South Wales they took, by number, about 20 per cent Sugar Gliders, 20 per cent rabbits, 19 per cent birds and 56 per cent insects (Debus *et al.* 2005). In a Queensland study, they took 25 per cent mammals by number, 71 per cent by prey biomass, mainly Sugar Gliders, birds – 10 per cent by number, 24 per cent by biomass and arthropods – 65 per cent by number, 1 per cent by biomass (Barnes *et al.* 2005). They use short-stay perch-hunting, flying or swooping to seize prey from the ground, air, trees and shrubs or from the surface of water. It is said to be the Australian owl most likely to catch food on or in the water, i.e. small Water Rats (*Hydromys chrysogaster*) and small ducks or grebes (Higgins 1999; Debus 2009).

Breeding: Barking Owls are said to form monogamous, lifelong pairs that are resident in large, permanent home ranges (Debus 2009), but this needs confirmation with decades-long studies of several marked pairs and territories. More likely, certain females change territories more often than males do and home range sizes change over time (Olsen and Trost 2007). Home range sizes are estimated at 200 to 1000 ha, but new data suggest that home ranges might be up to 5000 ha in some habitats (Debus 2009). The owl lays in winter and spring, usually from July or August to September or October, slightly earlier in the north than in the south (Higgins 1999). The nest is a large hollow up to 46 cm wide and 3 m deep in a big old tree, usually live but sometimes dead, up to 35 m above the ground. Exceptional sites include a crevice in the sheltered main fork of a tree, overhung by foliage; crevices or caverns in rocks; and perhaps rabbit burrows (Debus 2009).

Eggs: The clutch is one to three eggs, usually two, rarely four. Incubation takes 34–38 days.

Young: The nestling period is said to be about 5 weeks (Higgins 1999) but 5 weeks is probably the branching period, not fledging which may be closer to 6 weeks. Successful broods often have one, two or three young, an average of 1.9 young per successful brood in New South Wales. Young are dependent on their parents for several months, possibly for up to 5 months (Higgins 1999; Debus 2009).

Voice and calls: The Barking Owl's characteristic call is a rapid, gruff double hoot *wook-wook*, said to be higher pitched and more terse in the female. Each couplet occupies about half a second, and is repeated at intervals of a few seconds. At close range, the first note of the couplet is preceded by a brief, soft growl. The nature of the call is such that it can be difficult to decide whether it is a distant dog or an owl, and many false records of Barking Owls are attributable to dogs barking or foxes yapping. The owls commonly give the 'barking' call in a duet between members of a pair. Other calls include dog-like growling sounds, rising to snarling in defence of owlets, and a human-like wavering scream responsible for the folk name 'screaming woman owl'. The juvenile begging call is a grating trill, deeper and more drawn-out than that of the Southern Boobook. The Barking Owl's double 'bark' is more rapid than the leisurely double hoot of the Southern Boobook (Olsen *et al.* 2002a; Debus 2009).

Status: The Barking Owl is not globally or nationally threatened, being listed in Australia as Lower Risk – Least Concern (Garnett *et al.* 2011). However, the assumed subspecies *connivens* of eastern and southern Australia is considered Near Threatened, and is listed under state legislation as Vulnerable in New South Wales and South Australia and Endangered in Victoria. If Barking Owls in south-western Australia are a distinct subspecies, they deserve listing as Vulnerable or perhaps Endangered (Debus 2009).

Barking Owls are common in the Tropics, but uncommon to rare and declining in the south. In temperate Australia they are threatened by clearance and fragmentation of habitat, by intensive logging or firewood harvesting in inland forests, especially box woodland, by loss of old hollow trees on private land and perhaps by feral Honeybees (*Apis mellifera*) occupying hollows. There might be 7500 breeding pairs of the subspecies *connivens* in eastern and southern Australia, mostly in Queensland, with perhaps a similar number of pairs in northern Australia (Debus 2009).

Myths and questions: Accurate fledging times, defined as young fully feathered or flying, need to be determined. The status of these 'subspecies' needs to be clarified. The designated subspecies do not seem to be based on clear evidence. This is a trans-equatorial, tropical owl species, common in northern Australia, its stronghold. South-eastern Australia is probably marginal habitat.

Southern Boobook *Ninox novaeseelandiae*

The species name *novaeseelandiae* is for 'New-Zealand Owl', the type locality.

Field recognition: The Southern Boobook is a medium-sized owl, 30–35 cm long with a wingspan of 56–82 cm. On mainland south-east Australia males weigh about 250 g, females about 315 g; they are smaller in Tasmania, where males average about 190 g and females about 214 g. They are a compact owl with a large head, short, broad, rounded wings, a short tail and are highly variable in plumage. The **adult** is light brown to rich dark brown on the upperparts, with white spots on the shoulders and barring in the wings and tail. The face and eyebrows are white, with a contrasting dark patch around each eye. The underparts are correspondingly light brown to brown, with white markings varying from streaks to mottling or coarse spotting. The palest form occurs in the inland and north (subspecies *ocellata*, 'spotted'), and the darkest form occurs in rainforest of northern Queensland (subspecies *lurida*, 'red'). Tasmanian birds (subspecies *leucopsis*, 'white face') are finely and profusely spotted white on the upperparts. The bill is black with a light blue-grey base, the cere is light blue-grey, the eyes are hazel on the mainland or yellow in Tasmania, and the feet are light grey (pink-brown in the darkest birds). The **juvenile** is much paler and downy on the head and underparts, with fine streaks on the breast and dark eye-patches (Higgins 1999; Debus 2009).

Subspecies: The taxonomy of Southern Boobooks was confusing back in the days of David Fleay (1968), and remains so today. König *et al.* (1999) considered New Zealand and Australian Boobooks to be separate species, *N. novaeseelandiae* and *N. boobook*, and listed four subspecies for *boobook* in Australia. In contrast, Higgins (1999) considered New Zealand and Australian boobooks to be the same species, and listed the same four subspecies for Australia – *boobook* (described above) from eastern and southern Australia to the base of Cape York Peninsula and west to about Port Augusta in South Australia, *lurida* a small dark form of rainforest in north-east Queensland, *ocellata* a pale, variable form widely distributed in north, central and western Australia, including Cape York Peninsula, and *leucopsis* a small spotted form in Tasmania. Two other forms, *undulata* from Norfolk Island and *albaria* on Lord Howe Island, are extinct. Olsen (1999) considered Tasmanian and New Zealand forms to be the same species, a different species from those on the Australian mainland. This has been discredited (Olsen and Debus 2005). Christidis and Boles (2008) note that although the DNA evidence indicates that Australian and New Zealand forms are best treated as one species, it is possible that other forms of the complex in the Timor region (Olsen *et al.* 2010), and the form *lurida* in the Australian Wet Tropics, could prove to be separate species. DNA sequence data are required to resolve these issues.

Christidis and Boles (2008) note that it is important to correct two major misquotations in the literature on the findings of Norman *et al.* (1998b). Olsen (1999) stated that molecular data of Norman *et al.* (1998b) supported the separation of the subspecies *boobook* and *novaeseelandiae* as separate species and the inclusion of Tasmanian *leucopsis* in *novaeseelandiae*, the New Zealand form. Higgins (1999) stated that Norman *et al.* (1998b) concluded that *novaeseelandiae*, *undulata* and *leucopsis* formed a group separate from those of mainland Australia. Both statements are in error. Norman *et al.* (1998b) did not present data on Australian mainland *boobook*. Instead they included *leucopsis* as a representative of the Australian forms for comparisons with extralimital forms, which leaves open the question of mainland Boobooks in relation to Tasmanian, New Zealand, Lord Howe and Norfolk Island Boobooks.

Boobook taxonomy was further complicated by the treatment in Wink *et al.* (2008). Using the cytochrome b gene they concluded that *boobook* (mainland

Figure A.8 Possible breeding range of the Southern Boobook in Australia.

Australia) and *novaeseelandiae* (New Zealand) were distinct species and that *leucopsis* in Tasmania was a third distinct species. If so, this has important implications for species conservation (see Chapter 40).

Range: The Southern Boobook occurs throughout mainland Australia, Tasmania and most major offshore islands (Figure A.8), as well as New Guinea and Timor, unless *fusca* is a full species (Olsen *et al.* 2010).

Migration: The Southern Boobook is said to be a partial migrant in Tasmania, some birds crossing Bass Strait in autumn. Individuals that look like the Tasmanian *leucopsis* subspecies have been recorded in Victoria and New South Wales in winter (Higgins 1999). However, it is not clear whether this is a regular migration, as opposed to dispersing individuals, and the evidence for Bass Strait migration needs confirmation with marked birds. Near Canberra some females leave the breeding territory and form winter home ranges (Olsen and Taylor 2001).

Habitat, habits and food: The Southern Boobook lives in open and closed habitats with trees, buildings or other shelter such as caves, cliffs or rocky areas. It roosts in foliage, tree hollows, other natural holes or crevices, caves and in buildings. Its flight is agile and goshawk-like, with bursts of quick wing-beats and glides; it is able to catch insects, bats and birds in mid-air (Debus 2009; J. Olsen and S. Trost unpublished).

Resident pairs of Southern Boobooks have traditional roosting sites, which can be close to or away from the nest depending on the time of year, stage of the breeding cycle, prey and relationships with neighbouring Boobooks (Olsen and Trost 1997, 2007). The owls usually roost alone, but sometimes in pairs or family groups. Roosts are used in rotation, from days to weeks at a time; some roosts are used irregularly. Adults, especially the female, may defend fledglings at night against humans and other animals (Olsen *et al.* 2002a).

At dusk Boobooks rouse, mantle and stretch. On dark some might advertise with territorial Boobook calls or maintain contact with bouts of *boo-book* calling. Neighbouring birds may respond. Females give territorial calls less often, but give them particularly in the pre-laying phase. They do not duet (Olsen *et al.* 2002b). At dawn the owls return to their roosts, sometimes arriving after first light and after the dawn chorus of diurnal birds.

In the courtship period, on rousing at dusk, males or females sometimes give bouts of *boo-book* calls. Courtship proceeds with long bouts of croaking calls (often at the nest hole), allopreening and *boo-book* calling by the male and female. The male provides food to the female, which begs with Braying calls. These and other calls accompany mating, on a branch near the nest, and culminate in a rabbit-like Mating Squeal (Olsen *et al.* 2002a; Debus 2009).

In response to a persistently intruding owl, as simulated by call playback, a defending Boobook will often approach and give croaking or *boo-book* calls. However, they may remain silent nearby and watch. Rival males can escalate calling into vocal Duels that sometimes culminate in fighting, with chittering calls (Olsen and Trost 1997). In the breeding season, pairs may respond to playback by sometimes copulating, with the female giving a Mating Squeal after giving croaking and *boo-book* calls (Olsen *et al.* 2002b). Boobooks sometimes call in response to playback of other *Ninox* calls, and will harass large owls such as the Powerful Owl. They sometimes stop calling if the calls of Masked or Powerful Owls are broadcast (Debus 2009; J. Olsen and S. Trost unpublished).

Southern Boobooks bathe after dusk or before dawn. However, the claims that they sunbathe by lying on the ground with their wings stretched out (Higgins 1999; Debus 2009) need confirmation. Boobooks are heavily mobbed if they move around in daylight, and these claims may refer to captives, i.e. Fleay (1968) which shows a photo of a Boobook sunning.

The Southern Boobook preys on a variety of mammals, birds, lizards, frogs and invertebrates. They take many vertebrates in winter and during the breeding cycle, but juveniles and non-territorial 'floaters' take proportionally more invertebrates compared to breeding adults. Boobooks in warmer regions probably take proportionally more invertebrates (Rose 1996; Olsen *et al.* 2006). Rose (1996) made 12 collections of pellets (not 12 pellets as stated in Higgins 1999), and found, by number, small marsupials in 12 per cent of collections, bats in 25 per cent, rodents in 58 per cent, birds in 42 per cent, beetles in 42 per cent and spiders in 92 per cent. Trost *et al.* (2008) found in winter diet near Canberra, 37.7 per cent mammals, 6.7 per cent birds and 59.7 per cent invertebrates, but by biomass (weight), Boobooks took 61.1 per cent mammals, 36.1 per cent birds and 2.8 per cent invertebrates.

Boobooks perch and watch for prey, swooping to snatch prey from the ground, off foliage or out of the air. They crash into foliage after insects, geckos and birds, and take birds, bats and insects in mid-air (J. Olsen and S. Trost unpublished data). They also take mice flushed by agricultural machinery and insects around artificial lights.

Breeding: Southern Boobooks form monogamous pairs during the breeding season in defended breeding territories of around 100 ha near Canberra, but probably larger in poor habitat and smaller in rich habitat. In southern Australia they tend to lay in spring, usually September–November, but sometimes in late winter (July–August) and extending to summer (December–February or later). The nest is a hollow up to 30 cm wide and can be 20 cm to the back of the nests and 0 to over 2 m down from the entrance in a live or dead tree. Other sites include cavities in buildings, old stick nests of crows, rock ledges or caves.

Eggs: The clutch is usually two or three eggs, but may be one to four, or exceptionally five. Incubation takes 29–31 days.

Young: The nestling period is about 5 weeks to branching and 6 weeks to fledging. In Australia, broods of nestlings average 2.8, and successful nests average 2.6 fledglings. Young are dependent for about 30 to 60 days after fledging, until their first autumn or winter (Higgins 1999; Debus 2009; Olsen and Trost unpublished data).

Voice and calls: The Southern Boobook's characteristic call is its territorial song, a high-pitched disyllabic hoot – *boo-book*, responsible for the folk names 'mopoke' or 'morepork'. Each couplet takes about half a second and is repeated about every 4 or 5 seconds, sometimes up to and beyond 1 hour. The second note of the couplet is lower pitched than the first. Neighbouring birds, especially males, duel. Other common calls include a long series of single or double guttural croaks (Por call and 'croaking *boobook*'), a rising, vibrating cry uttered by females (Braying), and both sexes give a Single Hoot when protecting fledglings from ground predators, including humans, and a soft purring call to young. The juvenile begging call is a shrill cricket-like trill, repeated (Olsen *et al.* 2002a).

Status: The Southern Boobook is not globally threatened, being listed in Australia as Lower Risk – Least Concern (Garnett *et al.* 2011). However, the subspecies *albaria* on Lord Howe Island and *undulata* on Norfolk Island are Extinct. These island subspecies suffered from loss of habitat, including nest sites, and possibly from competition or predation by introduced species. The Tasmanian subspecies is likely to suffer from competition for nest-hollows by feral Honeybees and introduced Laughing Kookaburras *Dacelo novaeguineae* and, if it turns out to be a separate species (Wink *et al.* 2008), conservation measures take on more urgency. The mainland Boobook is abundant, widespread and adaptable, occupying agricultural and urban landscapes. However, Barrett *et al.* (2007) calculated differences in reporting rates for birds in New South Wales between the first Atlas of Australian Birds, conducted between 1977 and 1981, and the New Atlas of Australian Birds conducted between 1998 and 2001. They found a 25 per cent decrease across New South Wales for Southern Boobooks *Ninox novaeseelandiae* but gave no reason for the decline. It has declined in some areas through loss of habitat and hollow trees, and is often hit by vehicles and sometimes collides with windows and wire fences (Debus 1999). Locally, brodifacoum-based rodenticides might have caused owl deaths as in New Zealand (Stephenson *et al.* 1999).

Myths and questions: The taxonomy of Southern Boobooks differs from publication to publication and needs further study and clarification. Many home range estimates are probably too small and this may affect assessments of their conservation needs, to the detriment of the species. Estimates need to be done with radio-telemetry in a variety of habitats. Fledging age needs to be clarified and differentiated from branching age. Finally, the claim that Tasmanian individuals migrate across Bass Strait needs to be verified.

Christmas Island Hawk-Owl *Ninox natalis*

The species name *natalis* is for Christmas Island, which was named on Christmas Day 1643.

Field recognition: The Christmas Island Hawk-Owl is a small owl, 26–29 cm long with a wingspan of 61–67 cm and weighing 140–200 g. The **adult** is mostly rufous-brown with a darker mask, bright yellow eyes and fine white barring and spotting; they are about two-thirds the size of a Southern Boobook. The **juvenile** is similar to the adult but slightly paler with downy head, neck and underparts. There are no other owls on Christmas Island (Higgins 1999). Though females are said to be slightly larger, the data are limited and Hollands (2008) thought some males were at least as big as females.

Subspecies: None. The Christmas Island Hawk-Owl was recently treated as a subspecies of *N. squamipila* the Moluccan Hawk-Owl. This species was said to comprise five nominal forms: *squamipila* (Seram), *hypogramma* (Halmahera, Ternate, Bacan), *hantu* (Buru), *forbesi* (Tanimbar) and *natalis* on Christmas Island. Norman *et al.* (1998a) examined the taxonomy of three of these nominate forms *squamipila*, *hypogramma* and *natalis* and concluded they did not form a monophyletic group relative to *N. connivens*, the Barking Owl, and *N. novaeseelandiae* the Southern Boobook. They concluded, from genetic distances

Figure A.9 Breeding range of Christmas Island Hawk-Owl.

and tree topography, at least three separate species in the complex: *N. natalis*, Christmas Island Hawk-Owl, *N. hypogramma*, the Northern Moluccan Hawk-Owl and *N. squamipila*, Southern Moluccan Hawk-Owl *contra* Olsen (1999) (Christidis and Boles 2008). More recently, a third member within the Moluccan complex has been recognised as a full species: the Tanimbar Hawk-Owl *N. forbesi* (Rheindt and Hutchison 2007).

Range: The Christmas Island Hawk-Owl is restricted to Christmas Island in the Indian Ocean, a 136.5 km^2 island 360 km south of Java, Indonesia (Figure A.9). Hill and Lill (1998a) found them on 46 of 158 of the 1 km^2 grid cells they surveyed on the island.

Migration: None known.

Habitat, habits and food: Christmas Island Hawk-Owls live in dense rainforest on both plateau and coastal areas. On the plateau they occur in tall evergreen forest including *Syzygium*, *Planchonella* and *Hernandia*. On the coast they use deciduous trees often with dense growths of epiphytic vines (Higgins 1999). They occasionally occur in disturbed or settled areas.

Hill and Lill (1998c) found that resident individuals or pairs, sometimes with young, day-roosted in the bottom third of the canopy. They preferred sheltered, concealed roosts.

Territory borders appeared to be fixed and defended through calling. In response to a persistently intruding owl, as simulated by call playback, a defending Hawk-Owl approached and gave croaking or *boo-book* calls. Rivals, especially males, could escalate this calling into vocal Duels at particular places along territory borders (Hill and Lill 1998a).

The Christmas Island Hawk-Owl hunts nocturnally. They prey mainly on large insects, such as Coleoptera (beetles) and Gryllacrididae (crickets), but also on spiders and small vertebrates including geckoes, birds and *Rattus* rodents. They watch from a perch, sometimes move from perch to perch, and swoop to snatch prey from the outer foliage of the canopy, a tree trunk, or occasionally from the air or the ground. They also sally forth from a perch and catch insects attracted to artificial lights around settlements, and they sometimes hunt along roadsides (Higgins 1999; Debus 2009).

Breeding: Reproduction in this species is poorly known. They appear to form monogamous, long-term pairs, which are resident in permanent home ranges of about 18 ha. The owl apparently has an extended breeding season, with fledged young noted in April, May and August–December (Higgins 1999). During courtship, the male provides food to the female, which begs with churring calls. A series of croaking notes, and allopreening, leads to mating, which takes place on a branch near the nest and culminates in a Mating Squeal (Higgins 1999; Debus 2009). Nest hollows have been found in live emergent forest trees, up to 18–26 m above the ground.

Eggs: The eggs have not been described, but a maximum clutch of two is likely. The incubation period is unknown but likely to be about 1 month (Higgins 1999; Debus 2009).

Young: The nestling period is not known. Subtracting the above possible incubation period (30 days) from the period given by Higgins (1999) for eggs and young in the nest (eggs laid to young fledged), 68 to 77 days minus 30 days, gives a possible nestling period of about 6 weeks (Debus 2009). Young are dependent on their parents for at least 2.5 months after fledging, and possibly longer (Higgins 1999).

Voice and calls: Hill and Lill (1998b) described six different calls for the Christmas Island Hawk-Owl similar to, but subtly different, from the calls of the Southern Boobook. Juveniles gave a begging *trill*, females gave a begging *chirring* call, males gave a *por-por* call during courtship and in territorial disputes, and a *kar-kar* call when delivering food. One or both sexes gave a *squeal* call during copulation and both sexes gave a disyllabic territorial *boo-book* call. Because of this call, Chinese workers on the island call this owl 'the dog which no man feeds'

(Higgins 1999). Hill and Lill (1998a,b,c) claimed that females had deeper voices than males, but this was not confirmed by the authors' electronic analysis and seems unlikely when other owls have deeper-voiced males compared with females, or the sexes are indistinguishable (Olsen *et al.* 2002a; Olsen and Trost 2003). It is not clear how Hill and Lill sexed live birds in the hand.

Status: The endemic Christmas Island Hawk-Owl is globally threatened, now listed as Endangered following recent concern over the proliferation of the introduced Yellow Crazy Ant (*Anoplolepis gracilipes*) on the island. Although present on the island for 70 or 80 years, this voracious ant has recently formed super-colonies that severely disrupt the surrounding forest structure, in turn affecting the owl's prey base and possibly preying on eggs and owlets. Recent interventions have successfully reduced numbers of these ants. The Hawk-Owl's limited population of about 560 pairs has been reduced by past clearance of about a quarter of its habitat for phosphate mining. In 1987 the owl's remaining habitat was protected within the Christmas Island National Park. However, part of the Park has since been annexed by the Australian Government for the construction of a detention centre for refugees, with mandatory environmental-impact assessment requirements waived 'in the national interest' (Garnett and Crowley 2000; Garnett *et al.* 2011; Debus 2009).

Myths and questions: It is unlikely that females are deeper-voiced than males, and it is not clear whether males or females are larger. There is no clear evidence as to how researchers sexed birds in the hand. There are different common names used for this species in different accounts – Christmas Hawk-Owl in Olsen (1999) but Christmas Island Hawk-Owl in König *et al.* (1999) and Higgins (1999). Christidis and Boles (2008) recommended Christmas Island Hawk-Owl.

References

Barnes CP, Rose AB and Debus SJS (2005) Breeding behaviour and diet of a family of Barking Owls *Ninox connivens* in south-eastern Queensland. *Australian Field Ornithology* **22**, 182–195.

Barrett GW, Silcocks AF, Cunningham R, Oliver DL, Weston MA and Baker J (2007) Comparison of atlas data to determine the conservation status of bird species in New South Wales, with emphasis on woodland-dependent species. *Australian Zoologist* **34**, 37–77.

Bunn DS, Warburton AB and Wilson RDS (1982) *The Barn Owl.* T & AD Poyser, Calton, UK.

Christidis L and Boles W (2008) *Systematics and Taxonomy of Australian Birds.* CSIRO Publishing, Melbourne.

Debus SJS (2009) *The Owls of Australia: A Guide to Australian Night Birds.* Envirobook, Sydney.

Debus SJS and Chafer CJ (1994) The Powerful Owl *Ninox strenua* in New South Wales. *Australian Birds* **28** (Supplement), 21–39.
Debus SJS, Olsen J and Rose AB (2004) Diet of the Barn Owl *Tyto alba* near Lake Frome in arid South Australia. *Corella* **28**, 40–42.
Debus SJS, Ford JA and Rose AB (2005) Breeding season diet of a pair of Barking Owls near Armidale, New South Wales. *Corella* **29**, 15–16.
del Hoyo J, Elliott A and Sargatal J (Eds) (1999) *Handbook of the Birds of the World. Vol. 5. Barn-owls to Hummingbirds.* Lynx Edicions, Barcelona.
Estbergs JA and Braithwaite RW (1985) The diet of the Rufous Owl near Cooinda in the Northern Territory. *Emu* **85**, 202–205.
Fleay D (1968) *Nightwatchmen of Bush and Plain.* Jacaranda, Milton, Qld.
Garnett ST and Crowley GM (2000) *The Action Plan for Australian Birds 2000.* Environment Australia, Canberra.
Garnett ST, Szabo JK and Dutson G (2011) *The Action Plan for Australian Birds 2010.* CSIRO Publishing, Melbourne.
Higgins PJ (Ed.) (1999) *Handbook of Australian, New Zealand and Antarctic Birds, Vol. 4.* Oxford University Press, Melbourne.
Hill FAR and Lill A (1998a) Density and total population estimates for the threatened Christmas Island Hawk-Owl *Ninox natalis. Emu* **98**, 209–220.
Hill FAR and Lill A (1998b) Vocalisations of the Christmas Island Hawk-Owl *Ninox natalis*: Individual variation in advertisement calls. *Emu* **98**, 221–226.
Hill FAR and Lill A (1998c) Diet and roost site characteristics of the Christmas Island Hawk-Owl *Ninox natalis. Emu* **98**, 227–233.
Hogan FE and Cooke R (2010) Insights into the breeding behaviour and dispersal of the Powerful Owl (*Ninox strenua*) through the collection of shed feathers. *Emu* **110**, 178–184.
Hollands D (1991) *Birds of the Night.* Reed, Sydney.
Hollands D (2008) *Owls, Frogmouths and Nightjars of Australia.* Bloomings Books, Melbourne.
Johnsgard PA (1990) Hawks, *Eagles and Falcons of North America.* Smithsonian Institution, Washington DC.
Jones M, Juhaeni D, Banjaransari H, Banham W, Lace L, Linsley MD and Marsden SJ (1994) 'The status, ecology and conservation of the forest birds and butterflies of Sumba, Indonesia. A report to the Indonesian Directorate General of Forest Protection & Nature Conservation, the Indonesian Institute of Sciences and BirdLife International'. Department of Biological Sciences, Manchester Metropolitan University.
Kavanagh RP (2002) Comparative diets of the Powerful Owl Ninox strenua, Sooty Owl *Tyto tenebricosa* and Masked Owl *Tyto novaehollandiae* in southeastern Australia. In *Ecology and Conservation of Owls.* (Eds I Newton, R Kavanagh, J Olsen and I Taylor) pp. 175–191. CSIRO Publishing, Melbourne.

König C and Weick W (2008) *Owls of the World.* 2nd edn. Christopher Helm, London.

König C, Weick W and Becking J (1999) *Owls. A Guide to the Owls of the World.* Pica Press, Sussex, UK.

McNabb EG (1996) Observations on the biology of the Powerful Owl *Ninox strenua* in southern Victoria. *Australian Bird Watcher* **16,** 267–295.

McNabb EG, Kavanagh RP and Craig SA (2007) Further observations on the breeding biology of the Powerful Owl *Ninox strenua* in south-eastern Australia. *Corella* **31**, 6–9.

Mees GF (1964) A revision of the Australian owls. *Zoology Verhandelingen* **65**, 1–62.

Milledge D, Bower H, Bower S and Carlile N (2010) Strangers in paradise: the Masked Owls of Lord Howe Island. *Wingspan* **20** (4), 26–29.

Mooney N (1997) Habitat and seasonality of Masked Owls nesting in Tasmania. In *Australian Raptor Studies II.* (Eds G Czechura and SJS Debus) pp. 34–39. Birds Australia Monograph 3. Birds Australia, Melbourne.

Newton I, Kavanagh R, Olsen J and Taylor I (Eds) (2002) *Ecology and Conservation of Owls.* CSIRO Publishing, Melbourne.

Norman JA, Christidis L, Joseph L, Slikas B and Alpers D (2002) Unravelling a biogeographical knot: the leapfrog distribution pattern of Australo-Papuan sooty owls (Strigiformes) and logrunners (Passeriformes). *Proceedings of the Royal Society of London, Series B, Biological Sciences* **269**, 2127–2133.

Norman JA, Christidis L, Westerman M and Hill RFA (1998a) Molecular data confirms the species status of the Christmas Island Hawk-Owl *Ninox natalis. Emu* **98**, 197–208.

Norman JA, Olsen PD and Christidis L (1998b) Molecular genetics confirms taxonomic status of the endangered Norfolk Island Boobook *Ninox novaeseelandiae undulata. Biological Conservation* **86**, 33–36.

Olsen J and Debus S (2005) A comment on some errors in the literature regarding Australian owls. *Corella* **29**, 97–98.

Olsen J and Rehwinkel R (1995) 'Peregrines and Powerful Owls in Namadgi and Tidbinbilla'. Report for the National Estates Grants Program, Canberra.

Olsen J and Taylor S (2001) Winter home range of an adult female Southern Boobook *Ninox novaeseelandiae* near Parliament House, Canberra. *Australian Bird Watcher* **19**, 109–114.

Olsen J and Trost S (1997) Territorial and nesting behaviour in Southern Boobooks (*Ninox novaeseelandiae*). In *Biology and Conservation of Owls of the Northern Hemisphere: 2nd International Symposium; 1997 February 5–9; Winnipeg, Manitoba*. (Eds JR Duncan, DH Johnson and TH Nicholls) pp. 308–313. Gen. Tech. Rep. NC-190. U.S. Department of Agriculture, Forest Service, North Central Forest Experiment Station, St Paul, Minnesota.

Olsen J and Trost S (2003) Do Southern Boobooks duet or duel? *Canberra Bird Notes* **28**, 45–52.

Olsen J and Trost S (2007) Duelling and nest failure in Southern Boobooks *Ninox novaeseelandiae. Australian Field Ornithology* **24**, 13–25.

Olsen J and Trost S (2009) Common Brushtail Possums *Trichosurus vulpecula* evicts nestling Southern Boobook *Ninox novaeseelandiae* from its nest hollow. *Australian Field Ornithology* **26**, 46–52.

Olsen J, Debus S and Trost S (2010) Is the Timor Southern Boobook a separate species? *Boobook* **28** (1), 10.

Olsen J, Trost S and Hayes G (2002a) Vocalisations used by Southern Boobooks *Ninox novaeseelandiae* in the Australian Capital Territory. In *Ecology and Conservation of Owls.* (Eds I Newton, R Kavanagh, J Olsen and I Taylor) pp. 305–319. CSIRO Publishing, Melbourne.

Olsen J, Marcot BG and Trost S (2002b) Do Southern Boobooks *Ninox novaeseelandiae* duet? In *Ecology and Conservation of Owls.* (Eds I Newton, R Kavanagh, J Olsen and I Taylor) pp. 320–328. CSIRO Publishing, Melbourne.

Olsen J, Trost S and Myers SD (2009) Owls on the island of Sumba, Indonesia. *Australian Field Ornithology* **26**, 2–14.

Olsen J, Fuentes E, Rose AB and Trost S (2006) Food and hunting of eight breeding raptors near Canberra, 1990–1994. *Australian Field Ornithology* **23**, 77–95.

Olsen J, Judge D, Trost S, Rose AB, Flowers G, McAuliffe J, Lawrence R and Maconachie M (2011) Powerful Owl *Ninox strenua* diet from two sites in the Australian Capital Territory. *Australian Field Ornithology* **28**, (in press).

Olsen P (1999) Genus *Ninox* (boobook owls). In *Handbook of the Birds of the World. Vol. 5. Barn-owls to Hummingbirds.* (Eds J del Hoyo, A Elliott and J Sargatal) pp. 230–237. Lynx Edicions, Barcelona.

Pavey CR, Smyth AK and Mathieson MT (1994) The breeding season diet of the Powerful Owl *Ninox strenua* at Brisbane, Queensland. *Emu* **94**, 278–284.

Pizzey G and Knight F (2007) *The Field Guide to the Birds of Australia.* 8th edn. Harper Collins, Sydney.

Rasmussen PC and Anderton JC (2005) *The Birds of South Asia: The Ripley Guide. Vol. 2.* Lynx Edicions, Barcelona.

Rheindt FE and Hutchinson RO (2007) A photoshot odyssey through the confused taxonomy of Seram and Buru (southern Moluccas). *Birding Asia* **7**, 18–38.

Rose AB (1996) Notes on the diet of the Southern Boobook *Ninox novaeseelandiae* in New South Wales. *Australian Bird Watcher* **16**, 339–343.

Schodde D and Mason I (1980) *Nocturnal Birds of Australia.* Lansdowne, Melbourne.

Simpson K and Day N (1999) *Field Guide to the Birds of Australia.* 6th edn. Penguin, Melbourne.

Soderquist T and Gibbons D (2007) Home-range of the Powerful Owl (*Ninox strenua*) in dry sclerophyll forest. *Emu* **107**, 177–184.

Stephenson BM, Minot EO and Armstrong DP (1999) Fate of Moreporks (*Ninox novaeseelandiae*) during a pest control operation on Mokoia Island, Lake Rotorua, North Island, New Zealand. *New Zealand Journal of Ecology* **23**, 233–240.

Trost S, Olsen J, Rose AB and Debus SJS (2008) Winter diet of Southern Boobooks *Ninox novaeseelandiae* in Canberra 1993–2005. *Corella* **32**, 66–70.

Wink M, Sauer-Gürth H and Fuchs M (2004) Phylogenetic relationships in owls based on nucleotide sequences of mitochondrial and nuclear marker genes. In *Raptors Worldwide.* (Eds RD Chancellor and B-U Meyburg) pp. 517–526. World Working Group on Birds of Prey and Owls, Berlin and MME, Birdlife, Budapest.

Wink M, Heidrich P, Sauer-Gurth H, Elsayed A-A and Gonzalez J (2008) Molecular phylogeny and systematics of owls (Strigiformes). In *Owls of the World.* (Eds C König and F Weick) pp. 42–63. Christopher Helm, London.

Wink M, Heidrich P, Sauer-Gurth H, Elsayed A-A and Gonzalez J (2009) Molecular phylogeny of owls (Strigiformes) inferred from DNA sequences of the mitochondrial cytochrome *b* and the nuclear *RAG-1* gene. In *Proceedings Fourth World Owl Conference Oct–Nov 2007, Groningen, The Netherlands.* (Eds DH Johnson, D Van Nieuwenhuyse and JR Duncan). *Ardea* **97**(4), 581–591.

Appendix B: Rehabilitating injured and orphaned owls

Introduction

Too few people in Australia are trained to care for injured and orphaned owls. Stress, inadequate food, improper housing and premature release kill many rescued owls. Kept in sub-standard housing, they are disturbed by humans or domestic pets, thrash back and forth against wire walls, break feathers and damage ceres. A repeatedly broken feather stops growing; a damaged cere stops the beak from growing. Dogs, inquisitive children and looming faces all strike terror into owls, and inadequate food stops them from growing or healing. We want these owls back in the wild, but they often continue their run of bad luck even after they are released. They are returned to carers, or die because of broken feathers, beaks, weak muscles or bones and lack of hunting experience.

Injured and orphaned owls will survive captivity and release if the carer understands that owls are adapted to an environment where they have a balanced diet, a wide choice of perches, freedom from disturbance, and access to sun, water and breeze. These are discussed below. There is also useful information in books such as *Falconry: Care, Captive Breeding and Conservation* by Jemima Parry-Jones (1988) and *Caring for Birds of Prey* by J. Olsen (1990).

Picking up injured or captive owls

Minimising stress to the owl is your first priority. Wild birds are often killed by shock rather than injuries they receive. If an adult owl allows you to approach it, it may be in shock or badly injured. Don't try to examine the bird by yourself, this may further stress it. Concentrate on how to secure the owl.

Owls are powerful animals. Boobooks have astounding power in their feet and talons and the larger owls have even more. When held by humans, owls can struggle, break feathers, hurt themselves or use their powerful beaks and claws to injure the humans trying to help them. By following the steps below, you can safely pick up an orphaned or injured owl:

1) Gently throw a towel or jacket over the owl. This should engage rather than entangle the owl's claws, and cover its head.
2) Place your hand in the middle of the owl's back and press gently forward and down so the legs extend backwards out under its tail. Encircle the body, including the wings and legs, by wrapping both hands around the owl's lower back just above where the tail feathers emerge.
3) Pick up the owl with its beak facing away from you, and carefully disengage the owl's claws from the jacket or towel.
4) Place the owl in a darkened cardboard box (see below in 'Transporting owls') perforated with holes along its sides near the bottom, lined with newspaper or a towel. Do not use straw or sawdust, and do not place water in the box. Do not try to feed the owl; keep the owl cool.

Examining owls

Captive owls brought to your attention need to be examined with the help of an experienced carer or a veterinarian. A veterinarian should carry out a complete examination, but carers can do a simple preliminary examination, starting from the head and working towards the tail. First, determine the species from a field guide, and whether it is an adult or a fledgling. It takes some experience to separate a wild from hand-raised owl because many hand-raised owls stay wild and mistrustful of humans, and may thrash around the cage when humans approach. Other hand-raised owls are tamer and vocalise, soliciting food from humans, something injured adult owls never do. Examine the wing and tail feathers, where they emerge from the body. If all of them are still growing, the bird is a fledgling, but if none or only a few wing or tail feathers are still growing, it is a full-grown young bird of that year or an adult. Fledgling *Ninox* owls look fluffy and downy on the head and body, fledgling *Tyto* owls have traces of down on their head and thighs. Owls, unlike chickens, are full-sized when they are fully feathered, soon after leaving the nest, at about 6 to 10 weeks depending on species.

The bird's head should be covered so the handler is not bitten, while the wings, legs and feet are stretched out for examination.

Head: Examine for injury or swelling.

Eyes: Examine for injury, quite common in owls. Shine a torch into the owl's eye in a darkened room to see if the pupil dilates. If it does not, neurological damage is indicated. If the eyelids are closing, the owl may have a disease or its condition may be low.

Cere and nostrils: Note injury to the cere, blockage or discharge from the nostrils. Blood may indicate cranial trauma.

Beak: Examine for fractures or splits, commonly caused by collision. Check inside the mouth for lesions (cheesy whitish growths may be caused by *Trichomoniasis*) or blood indicating internal injury. Listen for lung function; breathing should be quiet with no gurgling (fluid) or panting.

Neck: Check for injury. There is no point in checking for food in the crop. Owls have no crop.

Body: A sharp keel (*sternum*) indicates poor body condition. The preen gland at the base of the tail should appear normal, and feathers around it should be slightly oily. Blood around the vent may indicate internal injury; matting (excreta) around the vent indicates worms, poor health, disease or internal injury.

Wings and legs: Check for breaks or injury to bones or joints. Count the 10 primaries, then hold the body horizontal, with wings free, and rock the bird gently from side to side to check the symmetry of its wings and legs. The left wing should drop as you roll to the left, the right wing should drop as you roll to the right. Check for bumblefoot lesions, particularly on the pads of feet. These are areas of puffy convex skin, sometimes reddened and they indicate that an owl has been in captivity somewhere and forced to use a smooth round perch (injurious to raptors). Check for other foot injuries and feather loss.

Tail: Count the 12 tail feathers. Check for breakage and 'hunger traces' – these look like saw marks across the feathers and shafts. They cause weak points, and are generally produced by stress, starvation or bending the feather while the feather is still growing.

Keep the owl cool and calm during this process. Overheating and stress will kill a sick or traumatised owl. Dehydration is an extremely common cause of death in sick, injured or young owls. Even though many owls may get water from their prey in the natural state, it is crucial to leave a dish of water for the owl in its enclosure. Glucose-saline from the chemist will help re-hydrate a dehydrated owl.

Transporting owls

The general principles for transporting owls are (1) keep the bird cool and calm, and (2) prevent feather damage.

- The best container for transport is a darkened cardboard box or airline 'pet pack' with a hessian or towelling floor for the owl to grip, and big enough for the bird to turn around in without damaging its wings or tail. The best way to minimise stress, and keep the bird quiet, is to darken the box but punch holes low along its sides to allow ample circulation.

- Owls should not be transported during the hot part of the day (over 30°C) or in hot airless vehicles. Airliners are generally cool but the owl can become overheated while waiting to be placed on the aircraft. Though owls overheat quickly, low temperatures are no problem if they are fully grown and healthy.
- Don't give food or water to adults during trips of 24 hours or less.
- Growing talons and toes of nestling owls can be easily damaged. An owl's talons should be carefully dislodged from the hessian or towelling bottom of the container. To prevent its escape, remove the owl from the box inside the destination aviary or room. If an owl escapes, it will fly into glass and damage its head or neck and die or become permanently paralysed. Curtain all windows in the room, and close all doors.

Diet

The general principles when acquiring food for captive owls are to (1) make it fresh, (2) provide variety, and (3) make it is as similar as possible to the food owls eat in the wild. Owls need to process bones, fur, feathers and insect casings to produce castings (pellets). A good rule is to feed them enough so there is a little left over each day. Feed them once a day (three times a day for small orphans) and keep in mind that birds require more food in cold weather.

Chickens: Day-old cockerels contain egg yolk and are good food for owls. However, day-old chicks are too low in calcium to be a complete food. Some owl species fed exclusively on them will die, and growing owls will develop dietary deficiencies. Chickens 5–10 weeks old are a more suitable food.

Road-kills: Although you can feed fresh road-killed birds and other animals in the Australian winter, these go off quickly in the Australian summer, so take care.

Beef: In the first few days they can eat lean, fresh beef, but after that, where possible, use whole animals of an appropriate size for the owl, usually appropriate-sized birds or mammals. The addition of a pinch of natural bone meal or DCP (*dicalcium phosphate)* makes beef a more balanced diet, but if overused, both of these additives can kill owls.

Pigeons: Pigeons are highly nutritious, but they should never be fed to owls when fresh. Fresh pigeons carry *trichomoniasis* (frounce) and this kills raptors if the raptors are not medicated in the early stages. Freezing the pigeon seems to kill the protozoan *Trichomonas gallinae* that causes frounce.

Mice: Mice are a wholesome natural food for owls, and available from many universities.

Foods vary in nutritional value. The white meat of rabbits or chickens will not sustain a small owl like a Boobook. They need additional red meats like sparrows, beef or pigeons.

Orphaned owls

Unfortunately, most 'orphaned' young owls aren't orphaned at all. They are testing their wings and have fallen from the nest or roost. Just-branched owlets can flutter along the ground, even climb trees, but they cannot fly well. A grounded nestling may look abandoned, but the parents usually 'know' where it is. Nestlings will not be rejected because humans have touched them. If you place them in the branches of a tree close to where the owlet was found, parents usually come in and care for them. The carer should return to the area, wait until dark and listen for other trilling young, or for Bray calling or Purr calling of adult *Ninox*. Leave the owlet as high as possible in the tree and stand back far enough to allow adults to come in. In most cases, the juvenile will begin to Trill and the adult will 'rescue' it.

Fostering and hack

For those very few owlets that cannot go back to their own nest and parents, use **fostering** or **hack**.

Fostering

Fostering means placing an orphaned nestling into the nest with similar-aged young of the same species. This should work just as well for owls as it does for diurnal raptors (hawks, eagles, falcons), but there are fewer data on fostering owls. It needs further study. Some parent owls, unlike diurnal raptors, seem to recognise the calls of their own young, but there is no evidence that adult owls will harm fostered nestlings. Some foster parents may not feed them (Olsen *et al.* 2002), but McNabb (1994) successfully fostered Powerful Owls *Ninox strenua* back into the wild, so *Ninox* owls will accept foster young under certain circumstances.

Adult owls won't reject nestlings handled by humans, but they may reject sick or weak nestlings and nestlings that won't feed. Nestling owls should be placed in a foster nest when they are about half-grown (half-feathers/half-down) and eating well. If too old, they may leap from the nest when the climber places them there.

Hack

A nestling owl that has no hunting experience will probably starve if released on its own. *Ninox* owls are different from some diurnal (daylight) raptors because they often follow their parents from place to place inside the territory, being fed by the parents, and they roost during the day in different places. In contrast, fledged diurnal raptors, like falcons, goshawks or eagles return to a central place where parents deliver food, often to the nest, and don't move around and roost in different places as some owls do. Owls do not necessarily need their parents to teach them to hunt, but they do need practice in catching prey. **Hack** is a technique used to give young owls this hunting experience, to develop independence and strength, so they are not dependent on humans for food. This ancient method, used by falconers, involves the use of a hack-room or hack-box as an intermediate stage in the release process.

A hack box can be 0.5 m × 0.5 m × 0.5 m with a large hole 0.25 m × 0.25 m at the front covered with vertical bars so the owls can see out but can't get out. The box is placed high up in a bushy tree, shed, barn or shelter that fledged owls can later roost in during the day and feel secure. They are fed every day on a ledge or board at the entrance hole of the box so, when the bars are lifted and they are free, they know to come back to this ledge for food. In some cases, you can cut a hole in the back wall, covered by a flap, and place food into the rear of the hack-shed or box without being seen. This should prevent the owlets from flushing.

Nestlings are placed in the box about two weeks before they branch, when they are about one-third feathers and two-thirds down. When they are ready to branch (about 5 weeks with Boobooks, the juveniles commonly brought in) the bars in front of the hole are lifted during the day without alarming the nestlings and causing them to jump. Two days' supply of frozen food is tied to the food ledge and humans do not visit the hack box for two days.

Owls at hack are like wild owls. If humans come too close just before their initial flight, they can prematurely fledge and fly just strongly enough to evade humans. These juveniles will have no adults to guide them back to the hack box and will probably starve.

Many owlets are calm during the day. You can place food about 3 or 4 m away and the juveniles will stay put. It is very dangerous to flush owlets during the day because kookaburras, currawongs and other enemies attack and injure them. These enemies tend to leave the owlets alone during the day, if the owlets remain still.

You can monitor the fledglings at night with a torch and binoculars. It's fun to watch them fly clumsily from branch to branch and follow each other from tree to tree. Owls become solitary later in their lives, but at this stage *Ninox* fledglings Trill to each other and are very sociable. You can enjoy their nightly flight practice, play and adventures until they drift off (disperse). Expect them to stay around and take food for about 50 to 70 days.

Stephen Debus (1997) gives a thorough account of hack for nestlings in the *Tyto* genus, Masked Owl *Tyto novaehollandiae.*

Caring for young owls

As mentioned above, hand-raising owls should be avoided. If a nestling comes to you, it should be returned to its parents or fostered (see earlier section). However, under special circumstances, such as when no wild adults are available for fostering, hand-raising is necessary to prepare them for hack. Special care must be taken if the nestling is to be raised and made suitable for release.

Handling

Inexperienced carers often make the mistake of over-handling young owls. You should handle nestlings only when it is necessary, and then only with clean hands. Feeding and other care should take place with the chick inside a cardboard box lined with paper towels.

A nestling that is tame or imprinted to humans, that is, sees humans as its parents, may find itself in trouble after it can fly. It will continue to beg food from humans, even attack them. It is important to avoid taming owls or imprinting them to humans. Two basic principles should be adhered to:

1) give the nestling as much association with its own species as possible, for example, house a nestling Boobook with an adult Boobook; and
2) never let the nestling associate food with humans.

Young hand-raised owls should be fed with a simple cardboard puppet head, resembling the adult owl head, folded over tweezers. Use the tweezers to place meat into the nestling's mouth. Also, to avoid being seen by the nestling while feeding it, the carer can cover him or herself (head included) with a formless sheet.

Feeding

A nestling's first meal should be bite-sized pieces of meat, preferably ground whole animals like sparrows or mice that are small, soft and easily digested. Use curved forceps or tweezers and a puppet to place food into the open beak. For small owlets, 'Chirring' softly in imitation of the parent's call, can elicit a food-begging response from the nestling. They will reach up with an open beak.

Continue to include small quantities of fur and feathers in the diet of young owls. When the nestling is old enough to stand strongly, begin leaving it with whole carcasses of young fresh rats, small birds or mice, split open to expose the flesh.

Flight pens for rehabilitation

Owls need access to wind, rain, sun and fresh air but they also need shelter from these elements if they choose, and protection from predators, like cats, dogs, currawongs, kookaburras and other owls. Housing should reduce stress from humans or other fear-inducing elements. With this in mind, we can consider some plans for flight pens in Australia (Figure B.1). These designs are partially open to the weather because Australian owls are far more likely to die from heat stress than cold. They do need cover to give them the option of avoiding rain or storms. However, direct rain appears to help keep their plumage healthy and helps the beak naturally flake so it doesn't overgrow.

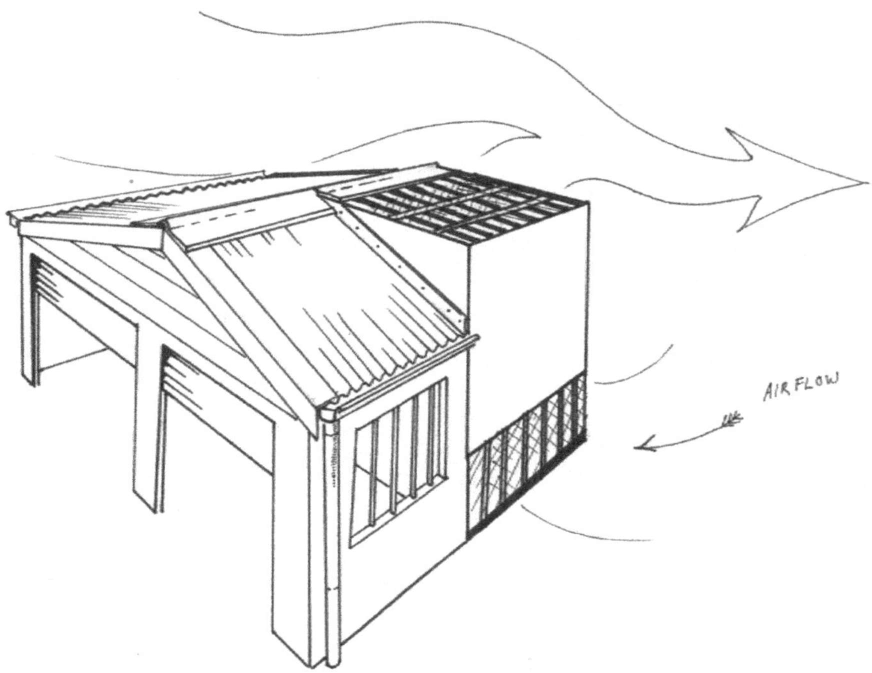

Figure B.1 Outside design of a flight pen.

General structure

Most owls ***cannot*** be kept behind wire. In wire cages they damage themselves, so use vertical (not horizontal) dowels or slats instead. (See Figures B.2 and B.3 for suitable pens.)

Doors: Owls are adept at flying out of a door when it is opened, so hang a safety curtain of canvas, or build a separate corridor outside the door.

Obstructions: Remove nails, loose pieces of string and any objects that could entangle or damage an owl.

Floor: Cover the floor with clean sand or pea-sized (1 cm) gravel. It is helpful if a small enclosure can be picked up or moved periodically to a clean area of ground.

Trees: It is useful to build the cage around an existing bushy tree, as David Fleay did (see Chapter 4), or plant a tree inside the cage. This gives the owls a shady place during the day. Perches up under the roof in one corner can also provide shelter.

Glass: Owls should **NEVER** be housed behind glass. In an attempt to escape, adults will fly into glass and damage their heads or necks, which can lead to the owl being permanently paralysed or even dying.

Perches: Remember that captive birds are perched more often and have fewer perch choices than wild birds do. Captive owls have poorer circulation in their feet because they are less active. Consequently, they are more prone to foot trouble like bumblefoot, a crippling infection in the pads of their feet. Some eucalypt branches, when used as perches, can damage an owl's feet because the wood is hard and the bark is eventually worn off, leaving a smooth rounded surface. Consequently, it is important to:

- provide a variety of perches. Include several that are wider than the talon span of the owl.
- use old vehicle tires, coconut fibre doormats, non-loop artificial doormats and artificial grass turf. These make excellent surfaces for perches. They prevent, and in some cases, cure bumblefoot. Coconut fibre mats are quite comfortable to owls and can be placed at an angle so perching birds don't soil them.

Bath/drinking pans: Bath pans need to be long enough for the owl to fit in, head to tail, but only deep enough to come up to the top of its tarsi. Plastic pigeon dishes work well. Bath pans 15 cm deep and 60 cm across are adequate for medium-sized owls. These must be kept filled with clean water, and away from perches so owls don't dirty the water with their own faeces.

Fire: Keep hessian bags and cardboard boxes available in case owls need to be evacuated from a bushfire.

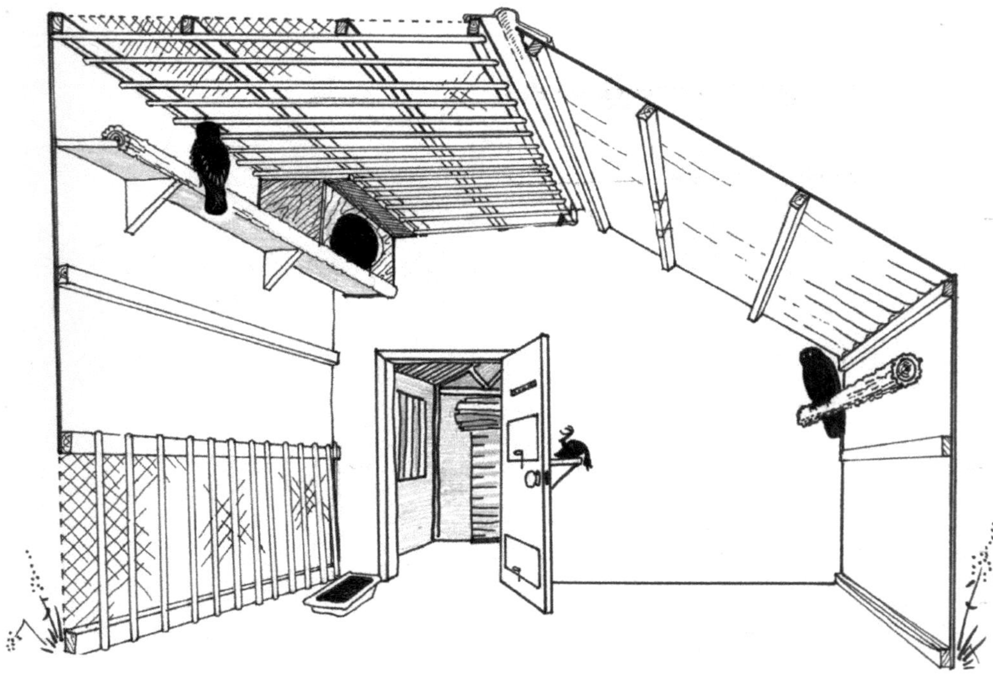

Figure B.2 Inside design of a flight pen.

Figure B.3 Injured Barn Owl in Colin McLaren's rehabilitation pen in Canberra.

Releasing owls

Determining when owls are ready for release

The important considerations for release are:

1) the health of the bird; and
2) the locality and time of release.

These two criteria appear straightforward but require considerable experience and assessment to judge.

Health of the bird

General

The first consideration should be general condition. If it was injured, the owl should be completely recovered and should have no broken feathers. The feathers should have a sheen on them and be waterproof. Talons and beak should be near

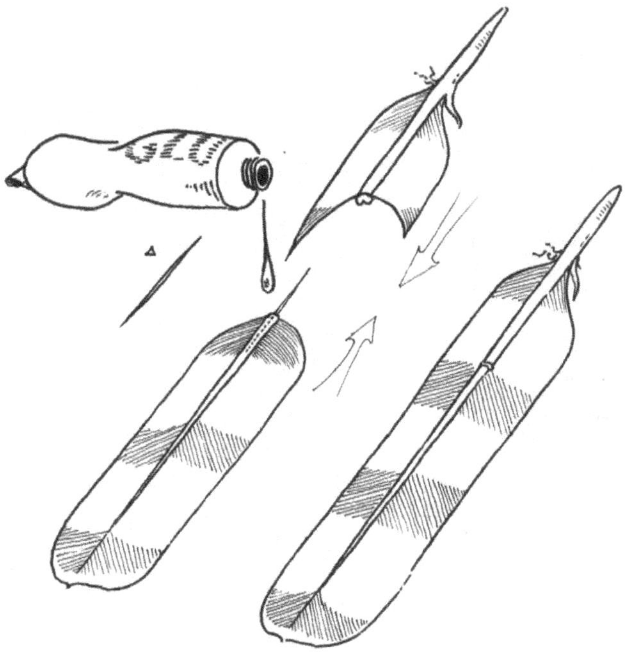

Figure B.4 Repairing a broken tail feather.

perfect and the bird should be in good condition, that is, muscle on either side of the sternum should be convex. This is best achieved by unlimited feeding while the individual is exercised in a large aviary.

Fixing broken feathers: 'Imping' can repair broken feathers. The process involves cutting the broken feather off cleanly and gluing in the same feather from another bird (see Figure B.4). Frank Beebe and Hal Webster discuss this process in detail in *North American Falconry and Hunting Hawks* (1964). See also Olsen (1994).

Exercise

After an injured owl has recovered, it can be exercised in an aviary by walking around inside the perimeter for 10 minutes a day, inducing the owl to fly during this time. It is best to increase this exercise gradually from 1 minute up to 10 minutes, as the owl's condition improves, with the owl flying most of the time during each session.

Flying ability

This can be difficult to ascertain. Birds flown using falconry techniques can be observed subjectively, or with the aid of video cameras, but wild owls are very difficult to train and fly. With most owls, flying in a large aviary, like the one described above, will give a reasonable indication of flying ability.

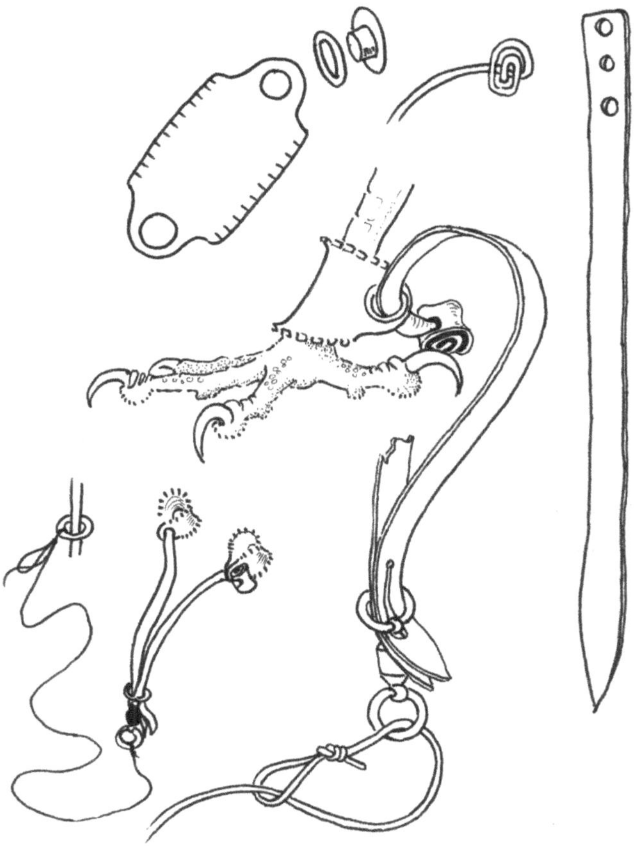

Figure B.5 Jesses, swivel and creance for testing flight ability.

If possible, two people can take the owl out on a piece of strong line and test the owl's flying ability. The bird is attached to the line by jesses and swivel (Olsen 1990) (see Figure B.5). One person releases the bird and the other lightly grips the line with gloves, providing a gentle brake that slows the bird to a stop after 30 m.

PLEASE NOTE: this technique can be dangerous – the line can break, or the bird can suddenly come to the end of the line and break a leg. Never tie the end of the line to a solid object. If the line slips from the gloved hands, a light block of wood dragged by the flying bird will slow it. Furthermore, currawongs, kookaburras and other enemies will attack owls by day, and an owl tethered by both feet cannot defend itself.

After the handler has experience with several releases, this technique gives a reasonable indication of flying ability, particularly if the owl is flown on the line twice upwind and twice downwind to judge its strength and symmetry. Examining the owl's flight on slow-motion video can help. However, the construction of a proper flight pen, as described above, should mean you won't have to test an owl's flying ability on a long piece of line.

Locality and time of release

Owl carers need to consult local ornithologists for help with release times and localities. Seasonal considerations are important. The dry season in Australia may be a better time for releasing most owls than the wet season because inclement weather prevents many owls from hunting. However, other factors are in play, particularly food availability.

With locality, the general principles to release owls are:

- in their natural habitat. For most species this means forest and woodland, though Grass Owls and Barn Owls can be released in open areas. Where possible, release them where they originally came from.
- where their natural food is abundant. Accessibility to food is important. Remember that high grass or heavy bush can make prey unavailable to some owls.
- away from territories of conspecifics (same owl species) or incompatible species. Mobbing from other birds is highly stressful to a newly released owl, so choose release sites with few potential mobbing birds.
- in suitable 'staging areas'. That is, localities where there are no breeding owls and few enemies. You can release owls in these places where they can remain for a time, undisturbed by humans, predators or other owls. There is abundant food but few nest sites in these staging areas (so territorial owls don't live there). We use two such areas successfully near Canberra, and you may find similar staging areas near your home.

Conclusion

Returning owls back to the wild is rewarding. Many carers view it as redressing the balance. After all, most owls were injured by humans or human-built structures in the first place. Rehabilitation puts us close to a wild creature for a time. We can nurture an adult or juvenile until release day, and then feel vicariously the sensation of freedom and independence that wild owls know.

References

Beebe F and Webster H (1964) *North American Falcons and Hunting Hawks.* Privately published, Denver, Colorado.

Debus SJS (1997) Aspects of biology of captive-bred, hack-released Masked Owls *Tyto novaehollandiae*. In *Australian Raptor Studies II.* (Eds G Czechura and S Debus) pp. 14–33. Birds Australia, Melbourne.

McNabb EG (1994) The successful rehabilitation of two Powerful Owl fledglings. *Australian Bird Watcher* **15**, 287–297.

Olsen J (1990) *Caring for Birds of Prey.* Applied Ecology Research Group, Canberra.

Olsen J (1994) *Some Time with Eagles and Falcons.* Applied Ecology Research Group, Canberra.

Olsen J, Trost S and Hayes G (2002) Vocalisations used by Southern Boobooks *Ninox novaeseelandiae* in the Australian Capital Territory. In *Ecology and Conservation of Owls.* (Eds I Newton, R Kavanagh. J Olsen, and I Taylor) pp. 305–319. CSIRO Publishing, Melbourne.

Parry-Jones J (1988) *Falconry: Care, Captive Breeding and Conservation.* David & Charles, London.

References

Barnes CP, Rose AB and Debus SJS (2005) Breeding behaviour and diet of a family of Barking Owls *Ninox connivens* in south-eastern Queensland. *Australian Field Ornithology* **22**, 182–195.

Barrett GW, Silcocks AF, Cunningham R, Oliver DL, Weston MA and Baker J (2007) Comparison of atlas data to determine the conservation status of bird species in New South Wales, with emphasis on woodland-dependent species. *Australian Zoologist* **34**, 37–77.

Beebe F and Webster H (1964) *American Falconry and Hunting Hawks.* Privately published, Denver, Colorado.

Bent AC (1938) *Life Histories of North American Birds of Prey Part 2.* Dover Publications, New York.

Bilney RJ, Cooke R and White J (2006) Change in diet of Sooty Owls (*Tyto tenebricosa*) since European settlement: From terrestrial to arboreal prey and increased overlap with Powerful Owls. *Wildlife Research* **33**, 17–24.

Botelho ES, Generro AL and Arrowood PC (1993) Parental care, nestling behaviors and nestling interactions in a Mississippi Kite *Ictinia mississippiensis* nest. *Journal of Raptor Research* **27**, 16–20.

Boyce DA (1985) Prairie Falcon prey in the Mojave Desert, California. *Journal of Raptor Research* **19**, 129–134.

Brouwer J and Garnett S (Eds) (1990) *Threatened Birds of Australia: An Annotated List.* Royal Australian Ornithologists Union and Australian National Parks and Wildlife Service, Moonee Ponds, Victoria.

Bull EL and Duncan JR (1993) Great Gray Owl (*Strix nebulosa*). In *Birds of North America, No. 41.* (Eds A Poole and F Gill) The Academy of Natural Sciences, Philadephia, and The American Ornithologists' Union, Washington DC.

Bunn DS, Warburton AB and Wilson RDS (1982) *The Barn Owl.* T & AD Poyser, Calton, UK.

Burton JA (Ed.) (1973) *Owls of the World: Their Evolution, Structure and Ecology.* EP Dutton, New York.

Campbell B and Lack E (1985) *A Dictionary of Birds*. T & AD Poyser, Calton, UK.

Caro TM and O'Doherty G (1999) On the use of surrogate species in conservation biology. *Conservation Biology* **13**, 805–814.

Chafer CJ (1992) Observations of the Powerful Owl *Ninox strenua* in the Illawarra and Shoalhaven Regions of New South Wales. *Australian Bird Watcher* **14**, 289–300.

Chitty D (1960) Population processes in the vole and their relevance to general theory. *Canadian Journal of Zoology* **38**, 99–113.

Christidis L and Boles W (2008) *Systematics and Taxonomy of Australian Birds*. CSIRO Publishing, Melbourne.

Cooke R, Wallis R, Hogan F, White J and Webster A (2006) The diet of Powerful Owls (*Ninox strenua*) and prey availability in a continuum of habitats from disturbed urban fringe to protected forest environments in south-eastern Australia. *Wildlife Research* **33**, 199–206.

Craighead JJ and Craighead FC (1956) *Hawks, Owls and Wildlife*. Dover, New York.

Davey SM (1993) Notes on the habitat of four Australian owl species. In *Australian Raptor Studies*. (Ed. P Olsen) pp. 86–100. Australian Raptor Association, RAOU, Melbourne.

Debus SJS (1996) Mating behaviour of the Southern Boobook *Ninox novaeseelandiae*. *Australian Bird Watcher* **16**, 300–301.

Debus SJS (1997) Aspects of the biology of captive-bred, hack-released Masked Owls *Tyto novaehollandiae*. In *Australian Raptor Studies II*. (Eds G Czechura and S Debus) pp. 14–33. Birds Australia Monograph 3, Birds Australia, Melbourne.

Debus SJS (2009) *The Owls of Australia: A Field Guide to Australian Night Birds*. Envirobook, Sydney.

Debus SJS and Chafer CJ (1994) The Powerful Owl *Ninox strenua* in New South Wales. *Australian Birds* **28** (Supplement), S21–S39.

Debus SJS, Ford JA and Rose AB (2005) Breeding-season diet of a pair of Barking Owls near Armidale, New South Wales. *Corella* **29**, 15–16.

Debus SJS, Olsen J and Rose AB (2004) Diet of the Barn Owl *Tyto alba* near Lake Frome in arid South Australia. *Corella* **28**, 40–42.

Delaney DK, Grubb TG and Beier P (1999) Activity patterns of nesting Mexican Spotted Owls. *Condor* **101**, 42–49.

del Hoyo J, Elliott A and Sargatal J (Eds) (1999) *Handbook of the Birds of the World. Vol. 5. Barn-Owls to Hummingbirds*. Lynx Edicions, Barcelona.

DeLong TR (1982) Effect of ambient conditions on nocturnal nest behaviour in Long-eared Owls. MSc thesis, Brigham Young University, Provo, Utah.

Dijkstra C, Bult A, Bijlsma S, Dann S, Meijer T and Zilstra M (1990) Brood size manipulations in the Kestrel *Falco tinnunculus*: effects on offspring and parent survival. *Journal of Animal Ecology* **59**, 269–286.

Downs JA and Horner MW (2008) Effects of point pattern shape on home range estimates. *Journal of Wildlife Management* **72**, 1113–1118.
Dwyer JF and Mannan RW (2007) Preventing raptor electrocutions in an urban environment. *Journal of Raptor Research* **41**, 259–267.
Earhart CM and Johnson NK (1970) Size dimorphism and food habits of North American owls. *Condor* **72**, 251–264.
Estbergs JA and Braithwaite RW (1985) The diet of the Rufous Owl near Cooinda in the Northern Territory. *Emu* **85**, 202–205.
Farabaugh SM (1982) The ecological and social significance of duetting. In *Acoustic Communication in Birds.* (Eds DE Kroodsma, EH Miller and H Ouellet) pp. 85–123. Academic Press, New York.
Flannery T (1994) *The Future Eaters.* Reed Books, Chatswood, New South Wales.
Fleay D (1968) *Nightwatchmen of Bush and Plain.* Jacaranda Press, Milton, Qld.
Fitzsimons JA and Rose AB (2007) The diet of a Southern Boobook *Ninox novaeseelandiae* in box-ironbark country, central Victoria. *Australian Zoologist* **34**, 85–88.
Gard NW and Bird DM (1990) Breeding behaviour of American Kestrels raising manipulated broods in years of varying prey abundance. *Wilson Bulletin* **102,** 605–614.
Garnett S (Ed.) (1992) 'Threatened and Extinct Birds of Australia'. RAOU Report 8. RAOU, Melbourne.
Garnett ST and Crowley GM (2000) *The Action Plan for Australian Birds 2000.* Environment Australia, Canberra.
Garnett ST, Szabo JK and Dutson G (2011) *The Action Plan for Australian Birds 2010.* CSIRO Publishing, Melbourne.
Gittleman JL, Funk SM, Macdonald D and Wayne RK (2001). *Carnivore Conservation.* Cambridge University Press, Cambridge, UK.
Glasier P (1978) *As the Falcon Her Bells.* Futura, London.
Hagen Y (1969) Norwegian studies on the reproduction of birds of prey and owls in relation to micro-rodent population fluctuations. *Fauna* **22**, 73–126
Higgins PJ (Ed.) (1999) *Handbook of Australian, New Zealand and Antarctic Birds, Vol. 4.* Oxford University Press, Melbourne.
Hill FAR and Lill A (1998a) Density and total population estimates for the threatened Christmas Island Hawk Owl *Ninox natalis. Emu* **98**, 209–220.
Hill FAR and Lill A (1998b) Vocalisations of the Christmas Island Hawk-Owl *Ninox natalis*: Individual variation in advertisement calls. *Emu* **98**, 221–226.
Hill FAR and Lill A (1998c) Diet and roost site characteristics of the Christmas Island Hawk-Owl *Ninox natalis. Emu* **98**, 227–233.
Hogan FE and Cooke R (2010) Insights into the breeding behaviour and dispersal of the Powerful Owl (*Ninox strenua*) through the collection of shed feathers. *Emu* **110**, 178–184.

Hollands D (1991) *Birds of the Night*. Reed, Sydney.
Hollands D (2008) *Owls, Frogmouths and Nightjars of Australia*. Bloomings Books, Melbourne.
Holthuijzen AMA (1990) Prey delivery, caching, and retrieval rates in nesting Prairie Falcons. *Condor* **92**, 475–484.
Johnsgard PA (1990) *Hawks, Eagles and Falcons of North America*. Smithsonian Institution, Washington DC.
Johnsgard PA (2002) *North American Owls: Biology and Natural History*. 2nd edn. Smithsonian Institution Press, Washington DC.
Jones M, Juhaeni D, Banjaransari H, Banham W, Lace L, Linsley MD and Marsden SJ (1994) 'The status, ecology and conservation of the forest birds and butterflies of Sumba, Indonesia'. Report to the Indonesian Directorate General of Forest Protection and Nature Conservation, the Indonesian Institute of Sciences and BirdLife International. Department of Biological Sciences, Manchester Metropolitan University, Manchester, UK.
Kavanagh RP (1988) The impact of predation by the Powerful Owl, *Ninox strenua*, on a population of the Greater Glider, *Petauroides volans*. *Australian Journal of Ecology* **13**, 445–450.
Kavanagh R (2002a) Comparative diets of the Powerful Owl *Ninox strenua*, Sooty Owl *Tyto tenebricosa* and Masked Owl *Tyto novaehollandiae* in southeastern Australia. In *Ecology and Conservation of Owls*. (Eds I Newton, R Kavanagh, J Olsen and I Taylor) pp. 175–191. CSIRO Publishing, Melbourne.
Kavanagh R (2002b) Conservation and management of large forest owls in southeast Australia. In *Ecology and Conservation of Owls*. (Eds I Newton, R Kavanagh, J Olsen and I Taylor) pp. 201–219. CSIRO Publishing, Melbourne.
Kavanagh R (2003) Conserving owls in Sydney's urban bushland: current status and requirements. In *Urban Wildlife: More than Meets the Eye*. (Eds D Lunney and S Burgin) pp. 93–108. Royal Zoological Society of New South Wales, Sydney.
Kavanagh RP and Bamkin KL (1995) Distribution of nocturnal forest birds and mammals in relation to the logging mosaic in south-eastern New South Wales, Australia. *Biological Conservation* **71**, 41–53.
Kavanagh RP and Jackson R (1997) Home-range, movements, habitat and diet of the Sooty Owl *Tyto tenebricosa* near Royal National Park, Sydney. In *Australian Raptor Studies II*. (Eds G Czechura and S Debus) pp. 2–13. Birds Australia Monograph 3, Birds Australia, Melbourne.
Kavanagh R and Murray M (1996) Home-range, habitat and behaviour of the Masked Owl *Tyto novaehollandiae* near Newcastle, New South Wales. *Emu* **96**, 250–257.
Kavanagh RP and Peake P (1993) Survey procedures for nocturnal forest birds: an evaluation of variability in census results due to temporal factors, weather and

technique. In *Australian Raptor Studies.* (Ed. P Olsen) pp. 86–100. Australian Raptor Association, RAOU, Melbourne.

König C and Weick W (2008) *Owls of the World.* 2nd edn. Christopher Helm, London.

König C, Weick W and Becking J (1999) *Owls. A Guide to the Owls of the World.* Pica Press, Sussex, UK.

Krebs C (2001) *Ecology: The Experimental Analysis of Distribution and Abundance.* Benjamin Cummings, San Francisco.

Krebs C (2008) *The Ecological World View.* CSIRO Publishing, Melbourne.

Lack D (1947) *Darwin's Finches.* Cambridge University Press, Cambridge.

Lack D (1954) *The Natural Regulation of Animal Numbers.* Oxford University Press, Oxford.

Lang AL, Duncan JR, Ramsay S and Rising JD (1991) Great Gray Owl and Northern Hawk Owl nests at Churchill, Manitoba. *Blue Jay* **49**, 208–214.

Lavazanian E, Wallis R and Webster A (1994) Diet of Powerful Owls (*Ninox strenua*) living near Melbourne, Victoria. *Wildlife Research* **21**, 643–646.

Leinonen A (1978) Hawk Owl breeding biology and behaviour at nest. *Lintumies* **13**, 13–18. (In Finnish with English summary.)

Liddelow GL, Wheeler IB and Kavanagh RP (2002) Owls in the southwest forests of Western Australia. In *Ecology and Conservation of Owls.* (Eds I Newton, R Kavanagh, J Olsen and I Taylor) pp. 233–241. CSIRO Publishing, Melbourne.

Linkhart BD and RT Reynolds (1987) Brood division and postnesting behavior of Flammulated Owls. *Wilson Bulletin* **99**, 240–243.

McKay GM (1983) Greater Glider. In *The Australian Museum Complete Book of Australian Mammals.* (Ed. R Strahan) pp. 134–135. Cornstalk Publishing, Sydney.

McNabb EG (1994) The successful rehabilitation of two Powerful Owl fledglings. *Australian Bird Watcher* **15**, 287–297.

McNabb EG (1996) Observations on the biology of the Powerful Owl *Ninox strenua* in southern Victoria. *Australian Bird Watcher* **16**, 267–295.

McNabb EG (2002) Notes on the diet and observations of the Southern Boobook (*Ninox novaeseelandiae*) in southern Victoria. In *Ecology and Conservation of Owls.* (Eds I Newton, R Kavanagh, J Olsen and I Taylor) pp. 192–198. CSIRO Publishing, Melbourne.

McNabb EG and Greenwood J (2011) A Powerful Owl disperses into town and uses an artificial nest-box. *Australian Field Ornithology* **28**(2), 56–75.

McNabb EG, Kavanagh RP and Craig SA (2007) Further observations on the breeding biology of the Powerful Owl *Ninox strenua* in south-eastern Australia. *Corella* **31**, 6–9.

Marks JS (1984) Feeding ecology of breeding Long-eared Owls in southwestern Idaho. *Canadian Journal of Zoology* **62**, 1528–1533.

Marks JS, Cannings RJ and Mikkola H (1999) Family Strigidae (typical owls). In *Handbook of Birds of the World. Vol. 5. Barn Owls to Hummingbirds.* (Eds J del Hoyo, A Elliott and L Sargatal) pp. 76–151. Lynx Edicions, Barcelona.

Marti CD (1974) Feeding ecology of four sympatric owls. *Condor* **76**, 45–61.

Marti CD (1976) A review of prey selection by the Long-eared Owl. *Condor* **78**, 331–336.

Marti CD, Korpimäki E and Jaksić FM (1993) Trophic structure of raptor communities: A three-continent comparison and synthesis. In *Current Ornithology, Vol. 10.* (Ed. DM Power) pp. 47–137. Plenum Press, New York.

Mees GF (1964) A revision of the Australian owls. *Zoologische Verhandelingen* **65**, 1–62.

Meinertzhagen R (1959) *Pirates and Predators.* Oliver and Boyd, Edinburgh and London.

Mikkola H (1972) Hawk-owls and their prey in northern Europe. *British Birds* **65**, 453–460.

Mikkola H (1983) *Owls of Europe.* T & AD Poyser, Calton, UK.

Milledge D, Bower H, Bower S and Carlile N (2010) Strangers in paradise: the Masked Owls of Lord Howe Island. *Wingspan* **20**(4), 26–29.

Mooney N (1997) Habitat and seasonality of Masked Owls nesting in Tasmania. In *Australian Raptor Studies II.* (Eds G Czechura and SJS Debus) pp. 34–39. Birds Australia Monograph 3, Birds Australia, Melbourne.

Murphy RK (1997) Importance of prairie wetlands and avian prey to breeding Great Horned Owls (*Bubo virginianus*) in northwestern North Dakota. In *Biology and Conservation of Owls of the Northern Hemisphere: 2nd International Symposium.* 5–9 February 1997, Winnipeg, Manitoba. (Eds JR Duncan, DH Johnson and TH Nicholls) pp. 286–298. USDA Forest Service General Technical Report NC-190, St Paul, Minnesota.

Newton I (1979) *Population Ecology of Raptors.* Buteo Books, Vermillion, South Dakota.

Newton I (1986) *The Sparrowhawk.* T & AD Poyser, Calton, UK.

Newton I (1998) *Population Limitation in Birds.* Academic Press, San Diego.

Newton I, Kavanagh R, Olsen J and Taylor I (Eds) (2002) *Ecology and Conservation of Owls.* CSIRO Publishing, Melbourne.

Norman JA, Christidis L, Joseph L, Slikas B and Alpers D (2002) Unravelling a biogeographical knot: the leapfrog distribution pattern of Australo-Papuan sooty owls (Strigiformes) and logrunners (Passeriformes). *Proceedings of the Royal Society of London, Series B, Biological Sciences* **269**, 2127–2133.

Norman JA, Christidis L, Westerman M and Hill RFA (1998a) Molecular data confirms the species status of the Christmas Island Hawk-Owl *Ninox natalis. Emu* **98**, 197–208.

Norman JA, Olsen PD and Christidis L (1998b) Molecular genetics confirms taxonomic status of the endangered Norfolk Island Boobook *Ninox novaeseelandiae undulata*. *Biological Conservation* **86**, 33–36.

Olsen J (1990) *Caring for Birds of Prey*. Applied Ecology Research Group, University of Canberra, Canberra.

Olsen J (1992) 'Raptors in Namadgi, Canberra Nature Parks, the Murrumbidgee River Corridor, and Googong Foreshore'. Report to ACT Parks and Conservation Service.

Olsen J (1994) *Some Time with Eagles and Falcons*. Applied Ecology Research Group, University of Canberra, Canberra.

Olsen J and Debus S (2005) A comment on some errors in the literature regarding Australian owls. *Corella* **29**, 97–98.

Olsen J and Georges A (1993) Do Peregrine fledglings reach independence during peak abundance of their main prey? *Journal of Raptor Research* **27**, 149–153.

Olsen J and Rehwinkel R (1995) 'Peregrines and Powerful Owls in Namadgi and Tidbinbilla'. Report for the National Estates Grants Program, Canberra.

Olsen J and Taylor S (2001) Winter home range of an adult female Southern Boobook *Ninox novaeseelandiae* near Parliament House, Canberra. *Australian Bird Watcher* **19**, 109–114.

Olsen J and Trost S (1997) Territorial and nesting behaviour in Southern Boobooks (*Ninox novaeseelandiae*). In *Biology and Conservation of Owls of the Northern Hemisphere: 2nd International Symposium*. 5–9 February 1997, Winnipeg, Manitoba. (Eds JR Duncan, DH Johnson and TH Nicholls) pp. 308–313. USDA Forest Service General Technical Report NC-190, St Paul, Minnesota.

Olsen J and Trost S (2003) Do Southern Boobooks duet or duel? *Canberra Bird Notes* **28**, 45–52.

Olsen J and Trost S (2007) Duelling and nest failure in Southern Boobooks *Ninox novaeseelandiae*. *Australian Field Ornithology* **24**, 13–25.

Olsen J and Trost S (2009) Common Brushtail Possum *Trichosurus vulpecula* evicts nestling Southern Boobook *Ninox novaeseelandiae* from its nest hollow. *Australian Field Ornithology* **26**, 46–52.

Olsen J and Tucker AD (2003) A brood size manipulation experiment with five nests of Peregrine Falcons *Falco peregrinus* near Canberra. *Emu* **103**, 127–132.

Olsen J, Debus S and Trost S (2010) Is the Timor Southern Boobook a separate species? *Boobook* **28**, 10.

Olsen J, Downs J, Tucker T and Trost S (2011) Home-range size and territorial calling of the Southern Boobook (*Ninox novaeseelandiae*) in adjacent territories. *Journal of Raptor Research* **45**, 136–142.

Olsen J, Fuentes E, Rose AB and Trost S (2006) Food and hunting of eight breeding raptors near Canberra, 1990–1994. *Australian Field Ornithology* **23**, 77–95.

Olsen J, Judge D, Trost S, Rose AB, Flowers G, McAuliffe J, Lawrence R and Maconachie M (2011) Powerful Owl *Ninox strenua* diet from two sites in the Australian Capital Territory. *Australian Field Ornithology* **28**, in press.

Olsen J, Trost S and Hayes G (2002a) Vocalisations used by Southern Boobooks *Ninox novaeseelandiae* in and adjoining Aranda Bushland, Canberra, Australia. In *Ecology and Conservation of Owls.* (Eds I Newton, R Kavanagh. J Olsen and I Taylor) pp. 305–319. CSIRO Publishing, Melbourne.

Olsen J, Marcot BG and Trost S (2002b) Do Southern Boobooks *Ninox novaseelandiae* duet? In *Ecology and Conservation of Owls.* (Eds I Newton, R Kavanagh, J Olsen and I Taylor) pp. 320–328. CSIRO Publishing, Melbourne.

Olsen J, Wink M, Sauer-Gürth H and Trost S (2002c) A new *Ninox* owl from Sumba, Indonesia. *Emu* **102**, 223–231.

Olsen J, Trost S and Myers SD (2009) Owls on the Island of Sumba, Indonesia. *Australian Field Ornithology* **26**, 2–14.

Olsen P (1999) Genus *Ninox* (boobook owls). In *Handbook of the Birds of the World. Vol. 5. Barn-owls to Hummingbirds.* (Eds J del Hoyo, A Elliott and J Sargatal) pp. 230–237. Lynx Edicions, Barcelona.

Parry-Jones J (1988) *Falconry: Care, Captive Breeding and Conservation.* David & Charles, London.

Pavey CR, Smyth AK and Mathieson MT (1994) The breeding season diet of the Powerful Owl *Ninox strenua* at Brisbane, Queensland. *Emu* **94**, 278–284.

Pearson TG (Editor-in-Chief) (1936) *Birds of America.* Garden City Publishing, Garden City, New York.

Penck M and Queale LF (2002) The diet of Southern Boobooks in South Australia. *Emu* **102**, 355–360.

Pizzey G and Knight F (2007) *The Field Guide to the Birds of Australia.* 8th edn. HarperCollins, Sydney.

Rabinowitz D (1981) Seven forms of rarity. In *The Biological Aspects of Rare Plant Conservation.* (Ed. H Synge) pp. 205–217. Wiley, Chichester, UK.

Rasmussen PC and Anderton JC (2005) *The Birds of South Asia: The Ripley Guide. Vol. 2.* Lynx Edicions, Barcelona.

Ratcliffe D (1993) *The Peregrine Falcon.* 2nd edn. T & AD Poyser, Calton, UK.

Ray JC (2005) Large carnivorous animals as tools for conserving biodiversity: assumptions and uncertainties. In *Large Carnivores and the Conservation of Biodiversity.* (Eds JC Ray, K Redford, R Steneck and J Berger) pp. 34–56. Island Press, Washington DC.

Ray JC, Redford KH, Steneck RS and Berger J (Eds) (2005) *Large Carnivores and the Conservation of Biodiversity.* Island Press, Washington DC.

Redpath SM, Leckie FM, Arroyo B, Amar A and Thirgood SJ (2006) Compensating for the costs of polygany in Hen Harriers *Circus cyaneus. Behavioral Ecology and Sociobiology* **60**, 386–391.

Rheindt FE and Hutchinson RO (2007) A photoshot odyssey through the confused taxonomy of Seram and Buru (southern Moluccas). *Birding Asia* 7, 18–38.

Riegert J and Fuchs R (2004) Insects in the diet of urban Kestrels from central Europe: an alternative prey of constant component of the diet? *Ornis Fennica* **81**, 23–32.

Roberson D (2000) Review of *Handbook of the Birds of the World, Vol. 5. Birding* **32**, 468–470.

Rohner C and Smith JN (1996) Brood size manipulations in Great Horned Owls *Bubo virginianus*: are predators food limited at the peak of prey cycles? *Ibis* **138**, 236–242.

Rohner C, Smith JM, Stroman J, Joyce M, Doyle FI and Boonstra R (1995) Northern Hawk-Owls in the Nearctic Boreal Forest: prey selection and population consequences of multiple prey cycles. *Condor* **97**, 208–220.

Rose AB (1993) Notes on the Powerful Owl in New South Wales. *Australian Birds* **26**, 134–136.

Rose AB (1996) Notes on the diet of the Southern Boobook *Ninox novaeseelandiae* in New South Wales. *Australian Bird Watcher* **16**, 339–343.

Schodde D and Mason I (1980) *Nocturnal Birds of Australia*. Lansdowne, Melbourne.

Sedgewick EH and Morrison PC (1949) Observations on the lower Murchison RAOU Camp, September 1948. *Emu* **48**, 212–242.

Sergio F, Marchesi L and Pedrini P (2004) Integrating individual habitat choices and regional distribution of a biodiversity indicator and top predator. *Journal of Biogeography* **31**, 619–628.

Sergio F, Newton I and Marchesi L (2008) Top predators and biodiversity: much debate, few data. *Journal of Applied Ecology* **45**, 992–999.

Sergio F, Newton I, Marchesi L and Pedrini P (2006) Ecologically justified charisma: preservation of top predators delivers biodiversity conservation. *Journal of Applied Ecology* **43**, 1049–1055.

Shields J (1994) Southern Boobook. In *Cuckoos, Nightbirds & Kingfishers of Australia*. (Ed. R Strahan) p. 130. Angus & Robertson, Sydney.

Simon S and Simon D (1980) Observation sur la reproduction de la Chouette Eperviere *Surnia ulula*. *Alauda* **48**, 152–153.

Simpson K and Day N (1999) *Field Guide to the Birds of Australia*. 6th edn. Penguin, Melbourne.

Smith HS (1935) The role of biotic factors in the determination of population densities. *Journal of Economic Entomology* **28**, 873–898.

Soderquist T and Gibbons D (2007) Home-range of the Powerful Owl (*Ninox strenua*) in dry sclerophyll forest. *Emu* **107**, 177–184.

Stephenson BM (1998) The ecology and breeding biology of morepork, *Ninox novaeseelandiae*, and the risk from secondary poisoning, in New

Zealand. Unpublished MSc thesis. Massey University, Palmerston North, New Zealand.

Stephenson BM, Minot EO and Armstrong DP (1999) Fate of Moreporks (*Ninox novaeseelandiae*) during a pest control operation on Mokoia Island, Lake Rotorua, North Island, New Zealand. *New Zealand Journal of Ecology* **23**, 233–240.

Tilley S (1982) The diet of the Powerful Owl *Ninox strenua* in Victoria. *Australian Wildlife Research* **9**, 157–175.

Tishechkin AK (1997) Comparative food niche analysis of *Strix* owls in Belarus. In *Biology and Conservation of Owls of the Northern Hemisphere: 2nd International Symposium.* 5–9 February 1979, Winnipeg, Manitoba. (Eds JR Duncan, DH Johnson and TH Nicholls) pp. 456–460. USDA Forest Service General Technical Report NC-190, St Paul, Minnesota.

Traill BJ (1993) The diet and movements of a pair of Powerful Owls *Ninox strenua* in dry eucalypt forest. In *Australian Raptor Studies.* (Ed. P Olsen) pp. 155–159. Australasian Raptor Association, RAOU, Melbourne.

Trost S, Olsen J, Rose AB and Debus SJS (2008) Winter diet of Southern Boobooks *Ninox novaeseelandiae* in Canberra 1997–2005. *Corella* **32**, 66–70.

Ulmschneider H (1990) Post-nesting ecology of the Long-eared Owl (*Asio otus*) in southwestern Idaho. MSc thesis, Boise State University, Boise, Idaho.

van Oosterzee P (1997) *Where Worlds Collide: The Wallace Line.* Reed, Victoria.

Wink M, Heidrich P, Sauer-Gurth H, Elsayed Abdel-Aziz and Gonzalez J (2008) Molecular phylogeny and systematics of owls (Strigiformes). In *Owls of the World.* (Eds C König and F Weick) pp. 42–63. Christopher Helm, London.

Wink M, Heidrich P, Sauer-Gurth H, Elsayed Abdel-Aziz and Gonzalez J (2009) Molecular phylogeny of owls (Strigiformes) inferred from DNA sequences of the mitochondrial cytochrome *b* and the nuclear *RAG-1* gene. In *Proceedings of the Fourth World Owl Conference*, October–November 2007, Groningen, The Netherlands. (Eds DH Johnson, D Van Nieuwenhuyse and JR Duncan). *Ardea* **97**, 581–591.

Wink M, Sauer-Gürth H and Fuchs M (2004) Phylogenetic relationships in owls based on nucleotide sequences of mitochondrial and nuclear marker genes. In *Raptors Worldwide.* (Eds RD Chancellor and B-U Meyburg) pp. 517–526. World Working Group on Birds of Prey and Owls, Berlin and MME/BirdLife, Budapest.

Winter J (1986) Status, distribution, and ecology of Great Gray Owls in California. MSc thesis, San Francisco State University, San Francisco.

Winter J (1987) 'Prey ecology of Great Gray Owls in Yosemite National Park'. Paper presented at Symposium on Biology and Conservation of Northern Forest Owls, 3–7 February 1987, Winnipeg, Manitoba.

Index